PRAISE FOR *TRUST AND .*

"For over 100 years, Australian and American forces have fought together as close allies. This important book on Australian Army mission command experiences across the globe shows, again and again, what we can learn from each other as we enter the next 100 years of mateship."
— Nick Warner, AO PSM, Director General, National Intelligence

"As the officer entrusted with the codification of Mission Command for the British Army in the mid-1990s, I much looked forward to reading and reviewing this title. I was not disappointed. This work is a most valuable contribution to the study of mission command in an army that has now embodied this decentralized philosophy of command in both doctrine and practice."
— Mungo Melvin, Major General, British Army (Ret.)

Trust and Leadership provides valuable commentaries on command aspects of Australia's past wars and particularly its more recent operations. Perhaps more importantly, it provides much food for thought for military professionals in both the Australian and U.S. armies. Indeed, it should be required reading for unit commanders and officers attending command and staff colleges."
— Dr. David Horner, Official Historian, Australian Peacekeeping, Humanitarian and Post-Cold War Operations

"There is no question that *Trust and Leadership* should be a recommended read for Australian audiences as well as those close partners who will continue to work with the Australian Army on operations. Trust and Leadership is an equally valuable reference for any student of the military art in any nation that seeks to have a better understanding of command and the culture that shapes it."
— Acton Kilby, Colonel, Canadian Armed Forces

"Mission command is in fact terribly difficult to conceptualize, let alone execute on the ground. The authors of *Trust and Leadership* have accomplished the extraordinary by painting a clear picture of mission command and showing the reader exactly what it means by way of real-world case studies. Nowhere else has the idea of mission command been so honestly presented in one collection. *Trust and Leadership* is a must-read for leaders, historians, and strategic thinkers."

—J. "Lumpy" Lumbaca, Lieutenant Colonel, U.S. Army Special Forces (Ret.)

"Skillfully, *Trust and Leadership* enlightens the reader with regard to the nature of mission command that has made it a secret ingredient of many successful military operations. It therefore merits the complete attention of everyone interested in military studies or leadership in general. Beyond the military domain, anyone interested in the structure of organizations, their efficiency, and their ability to adapt to change will gain valuable insights from this book."

—Luc Pigeon, Defence Research and Development Canada (DRDC)

"*Trust and Leadership* is, without question, both an excellent work of historical scholarship and an essential read for officers wise enough to know that history's lessons are neither preceptive nor generic. Mastering mission command is a constant work-in-progress, but the case studies contained herein offer an invaluable resource."

—Dr. David Stahel, University of New South Wales, Canberra

"Although written from an Australian perspective, sufficient explanation and comparison with U.S. ideas of mission command make this volume useful to a wide-ranging American and Commonwealth readership. All in all, it is a superbly edited volume that is well-written, solidly researched, and tightly put together."
— Dr. Howard Coombs, Associate Chair War Studies,
Royal Military College of Canada

"For the military professional, *Trust and Leadership* is mandatory reading. Soldiers at all grades need to understand how mission command can be optimised so that they can manage the complexities of current and future wars. This book will also find a welcome place on the shelf of the serious student of Australia's military past to understand the method commanders used to achieve their objectives helps to explain how the Army wages wars."
— Dr. Albert Palazzo, Director of War Studies,
Australian Army Research Centre

"I found this book to be an outstanding resource for military historians interested in learning more about the history of the Australian Army from WWI through deployments and action in both Iraq and Afghanistan. We learn that trust is an essential element of mission command and this trust between the higher HQ leaders and their subordinate leaders is the key to establishing the philosophy of mission command."
— John M. Allison Sr, Lieutenant Colonel, U.S. Marines (Ret.)

TRUST AND LEADERSHIP

The Australian Army Approach to Mission Command

Edited by Russell W. Glenn

UNG
UNIVERSITY *of*
NORTH GEORGIA™
UNIVERSITY PRESS

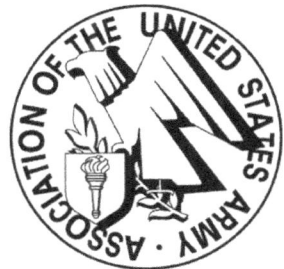

Published by:
University of North Georgia Press
Dahlonega, Georgia

Printing Support by:
Lightning Source Inc.
La Vergne, Tennessee

Cover and book design by Corey Parson.

ISBN: 978-1-940771-69-4

Printed in the United States of America

For more information, please visit: http://ung.edu/university-press
Or e-mail: ungpress@ung.edu

The views expressed in this book are the authors' and not necessarily those of the Australian Army, Australia's Department of Defence, Canadian government, or US Department of Defense. The Commonwealth of Australia will not be legally responsible in contract, tort, or otherwise for any statement made in this publication.

UNG
UNIVERSITY of
NORTH GEORGIA™
UNIVERSITY PRESS
Blue Ridge | Cumming | Dahlonega | Gainesville | Oconee

To the soldiers of the Australian Army, without
whom no mission would be accomplished

CONTENTS

FIGURES

FOREWORD

Although the concept of mission command is straightforward, employing it has been difficult for most Western militaries. Combat became fluid and dispersed years ago. With opportunities emerging and vanishing quickly, it has long been impossible for higher levels of command to direct subordinates' actions in detail. If fighting forces are to succeed, their commanders must entrust junior leaders with great freedom to act under the terms of a broad guiding intent. Yet, even though no other method of command works in the urgent press of combat or the need for immediate decisions in stability operations, the demands of Mission Command still trouble many commanders. A large number of them have resisted, restricted, or quietly rejected the idea.

Those leaders gravitate toward directive doctrines and detailed orders. There are cultural and institutional reasons for this. The stakes in operations are high and the preference for calculable outcomes grows naturally from both Western rationalism and military conservatism. Fears that the inexperience of junior officers will derail carefully planned actions or nullify the advantages of seasoned leadership add to the tendency to favor directive control.

There are risks to be considered when junior leaders' decisions change things fundamentally. In a well-known instance, Field Marshal Eric von Manstein, one of the leading proponents of mission command, saw one of his senior subordinates removed from command for using it too liberally. In the Crimean campaign of 1941, Lieutenant General Hans Graf von Sponeck, commanding 42nd Corps, deviated from Manstein's orders to defend the Kerch Peninsula. Even though he succeeded in meeting the overarching Army objective by stabilizing the defense after being outflanked, von Sponeck was court-martialed for falling back farther than the Army plan envisaged.

Despite the potential hazards for both superiors and subordinates, the chaotic nature of operations makes Mission Command and mission tactics indispensable. Most senior soldiers appreciate this, even if they do so grudgingly. Forces that extend trust to junior leaders must accept the risks of doing so. Carefully fashioned mutual understanding and cooperation mitigate the dangers of freeing subordinates to make crucial decisions and create the possibility for decisive outcomes both in combat and in stability operations.

Applying Mission Command emphatically does *not* mean delegating all authority to the lowest levels of command or refraining from intervening in operations as they progress. Senior commanders remain obligated to conceive clear, imaginative concepts that guide their operations to success. Applying necessary control and issuing essential detailed directives (as in coordinating one unit's movement through or around another or assuring that another agency's actions are optimally integrated) remains part of the commander's duty. Of primary importance is guaranteeing that all lower level leaders understand the broad intention that will govern action and shape initiative during execution.

Implementing this approach to command takes thoughtful effort from top to bottom of an organization. Mutual trust and constructive initiative have to be deliberately cultivated at every level to create the like-minded, mutually supportive and capable leader teams necessary. To succeed in this, the Prussians and Israelis, who have historically been the best practitioners of Mission Command, found it essential to shape their forces deliberately to implement the idea. They based *everything* on the demands of

Mission Command. Their recruitment and promotion, selection for command and staff assignments, and professional education and training all reflected the need for active, innovative leadership with a strong bias for recognizing opportunities and taking the initiative. Few other forces do that.

The essays in this book survey the practice of Mission Command in the Australian Army and offer insights about its application. Most of the authors are serving or retired military officers who discuss the subject in terms of their experiences during Australian and multinational efforts. The cases they present derive from operations in theaters spread from Iraq through Indonesia and Afghanistan to the Solomon Islands. Well-qualified teachers of command practice also contribute valuable perceptions and identify issues related to the subject.

Each chapter illuminates different aspects of Mission Command in its own way. Together they illustrate how applying the method differs substantially at tactical and operational levels. They also show how the actions of strategists and national or coalition authorities can limit or promote the effectiveness of Mission Command and how differences within international forces may enhance or impede it. Some of the writers relate how the opportunities and problems that arise from innovation in information technology and command tools can affect how a force is led and directed. These cases very usefully show the sometimes-unintended consequences of high-level guidance and intervention on commanders' freedom to act and to free their subordinates for independent action.

Australian forces are encountering these and other issues now. They will continue to do so as the national and international situations change and as military and civil-military cooperation matures. US and other military leaders face similar challenges. These essays will give readers from all those nations much to consider.

L. D. Holder
Lieutenant General (USA, retired)

FOREWORD

One thing I can assert about being a retired military officer is this: you never stop in your fascination for the study of the military arts and sciences. I have just finished a cover to cover reading of *Trust and Leadership*: *The Australian Army Approach to Mission Command*.

Let me start by congratulating all the contributors. I think it is an excellent analysis of the theory, histories, and case studies of the practice of mission command. In particular there is some quite pungent analysis of faulty mission commands; often extraordinarily valuable lessons may be drawn from bad practice as well as from good. Napoleon is hailed rightfully for his strong use of mission command but equally it might be observed that at the sunset of his military fame, at Waterloo in 1815, the failure of his subordinate Grouchy to keep Blucher from the field in the late afternoon was pivotal to Napoleon's defeat. Napoleon thought that Grouchy had his intent but was wrong in that assumption. That's the thing about mission command—often it is the crucial factor in spectacular success but from time to time for the sorts of reasons authors have outlined herein, it is the ingredient of failure or missed opportunity.

It is an uncomfortable feeling to find yourself quoted in a serious work of military science: "Mission command is essentially about professional trust between commanders and subordinates." Perhaps I should have added "and it works best when it rests within a framework of intent and limitations." Any military reader of this work will immediately contemplate their own opportunities and practice of mission command operations. Bob Hall, a Duntroon classmate of mine, very much describes the mission command culture at the junior levels of command in Vietnam. Company and platoon commanders had the broad intent of general officers, further expressed along with limitations, from their Commanding Officers, and thereafter the reins upon them were light.

Years later, in the International Force East Timor (INTERFET) campaign, the "commander's intent" from a combination of the UN, the various national political/military leaderships (including that of Australia) was essentially "to do needful things" to restore peace and security and to enable the further operations of the UN on the ground. There needed to be trust and trust is always engendered and increased by transparency. This issue of transparency is mentioned from time to time in the excellent chapters beneath. But I feel that mission command thrives when the senior commander is aware of any significant deviation from the anticipated plan, together with reasons. We weren't perfect in INTERFET but we avoided egregious challenges to the trust extended to us.

Another remark before allowing you to explore this fascinating read: one of the chapter authors laments opaqueness that diminished the potential effectiveness of his combat unit. He echoes the frustration of so many unit commanders before him and no doubt many who will follow. It's properly pertinent, therefore, to remark that there will always be layer upon layer of intent. Some will be absolutely obvious, contained in written instructions and available for enquiry and challenge. Other intent may be less obvious or even invisible to junior commanders. These facts of life are inimical to mission command; how can the junior commander feel like they know the score if there are believed to be matters of intent held in private. There is no easy answer and commanders must not roll up into a ball of frustration but grab every opportunity for local initiative.

Often though, the cleverest commanders will seize upon the jewel among the broken glass of all the other words. Lieutenant General Frewen, in his excellent chapter on the Solomon Islands matter ponders why, he as commanding officer of 2 RAR and thus junior to all the aspirant Colonels, et cetera, who wanted to lead the military part of the mission, was given the command by me; the mission command limitation that he described is quite accurate if a little vulgar. The mission command intent he quotes in the chapter, that what he would do would "set the face of ADF operations in the Pacific for decades to come" is correct. He understood that what he did with his force was very important but the way he did it was overwhelmingly high in my intent. That's why you got the job, General!

Trust, leadership, and transparency!

General Sir Peter John Cosgrove, AK, CVO, MC
26th Governor-General of Australia

1

MISSION COMMAND OVERVIEW

Dr. Russell W. Glenn

I do not propose to lay down for you a plan of campaign…but simply to lay down the work it is desirable to have done and leave you to execute it in your own way. Submit to me, however, as early as you can, your plan of operations.[1]

General Ulysses S. Grant to Major General William T. Sherman
April 4, 1864

Military operations, whether involving combat or otherwise, are complex and unpredictable. Intelligence, knowledge of one's own capabilities, and carefully crafted guidance at best lend limited insights into how to confront what lies ahead. Adversaries seek to deceive and surprise. Environmental conditions change. A wise military leader recognizes unforeseeable events always lie ahead. Commanders therefore require that subordinates adapt when confronted with the unexpected. Leaders' understanding of circumstances at the sharp end increasingly dims as one scales the chain of command even in this era of communications capabilities undreamt of a generation ago. The sergeant leading his squad sees what his platoon leader or company commander cannot. Those at battalion, brigade, and higher echelons know not what confronts each subordinate leader. Mission command—the practice of assigning a subordinate commander a mission without specifying how the mission is to be achieved—provides a means of addressing this challenge.[2] Centuries old in concept and decades aged in US Army doctrine, implementation nonetheless proves elusive. Fortunately, the United States is not the only country committed to mission command. Militaries in Australia, Canada, Germany, the Netherlands, New Zealand, Singapore, and United Kingdom are among those having adopted the philosophy in familiar form.

It is on Australia's approach that this book focuses. Australia has long been and continues to be a US ally and coalition partner of consequence. The two countries' soldiers served side-by-side in East Timor, Iraq, Afghanistan, on World War II battlefields, and elsewhere. There is great value in learning from those akin to but different from ourselves. The following pages should thereby prove insightful as America confronts future challenges to its security at home and abroad.

The authors offering insights include highly regarded academics and both serving and retired Australian Army officers. The academics take on earlier conflicts: World Wars I and II and that in Korea. All others were part of the events they consider. Any broader ruminations therefore have first-hand recollections in accompaniment, recollections that are at times quite unsparing. Those events include both confrontations with armed foes distant from Australian shores and disaster on the island continent.

US AND AUSTRALIAN PERSPECTIVES ON MISSION COMMAND

An order should not trespass upon the province of a subordinate. It should contain everything that the subordinate must know to carry out his mission, but nothing more…. Above all, it must be adapted to the circumstances under which it will be received and executed.[3]

US Army Field Manual 100-5,
Tentative Field Service Regulations, Operations (1939)

US and Australian views on mission command are similar in concept and in terms of the two countries' expectations regarding what the philosophy requires of senior and subordinate leaders alike. Not formally introduced as a term into US Army doctrine until 2003, the quotation above makes it clear that mission command has long been with America's army conceptually.[4]

US Perspectives on Mission Command

It is my design, if the enemy keep quiet and allow me to take the initiative in the spring campaign, to work all parts of the army together, and somewhat towards a common center…. You I propose to move against Johnston's army, to break it up and to get into the interior of the enemy's country as far as you can, inflicting all the damage you can against their war resources.[5]

General Ulysses S. Grant to Major General William T. Sherman
April 4, 1864

The US joint and army definitions of mission command are common in spirit but different in detail. Mission command in joint doctrine is "the conduct of military operations through decentralized execution based upon mission-type orders,"[6] The US Army as recently as 2018 defined the approach in rather stilted prose as

3

the exercise of authority and direction by the commander using mission orders to enable disciplined initiative within the commander's intent to empower agile and adaptive leaders in the conduct of unified land operations.... [It] emphasizes centralized intent and dispersed execution.[7]

In turn, disciplined initiative was "action in the absence of orders, when existing orders no longer fit the situation, or when unforeseen opportunities or threats arise."[8] The army recently saw fit to simplify its definition, describing mission command as the "approach to command and control that empowers subordinate decision making and decentralized execution appropriate to the situation."[9] Commendably streamlined in comparison to the previous description, we will see in the next section that it emulates the Australian Army's straightforward and readily grasped understanding of mission command.

Comprehensive employment of mission command continues to prove elusive across the entirety of the US armed forces.[10] Clear communication of a commander's intent is fundamental to subordinate understanding of what underlies an assigned mission. Intent—"a clear and concise expression of the purpose of the operation and the desired military end state [that] helps subordinate and supporting commanders to act...even when the operation does not unfold as planned"—allows junior leaders to act when confronted by the unforeseen.[11] An omniscient commander could provide precise instructions and the resources necessary for accomplishing every assigned task. No such commander has yet graced history, thus the need for providing subordinates with an intent to guide judgment when conditions vary from those envisioned. Stated at its simplest, an effective intent conveys what the commander wants his leaders and staff to remember when they face the unanticipated.[12]

The authority to act within the bounds of that intent is no less important than understanding it. The US Army still finds too many leaders practicing command characterized by tight control and overly detailed guidance. There are times when closer supervision is called for; effective leaders will judge when such is the case (of which more later). Finding the right balance between overly centralized control and an appropriately hands off approach depends on a number of factors, each requiring much from senior leaders. Perhaps that is why the default tends to be the former approach. The "nine thousand-mile screwdriver" representing Washington, D.C.'s overbearing control during the Vietnam War and echelons of command helicopters hovering over tactical engagements during that conflict lend the war a not undeserved reputation as one in which decentralized decision-making was granted too sparingly. The post-war 1982 and 1986 *Operations* manuals reminded a forgetful army that

> subordinates must act independently within the context of an overall plan. They must exploit successes boldly and take advantage of unforeseen opportunities. They must deviate from the expected course of battle without hesitation when opportunities arise to expedite the overall mission of the higher force. They will take risks, and the command must support them.[13]

Practicing such initiative allowed for the bold maneuver demonstrated by Colonel David Perkins with his 2003 "thunder runs" between Baghdad's international airport and the palace that would later become headquarters for coalition occupying forces.[14] Unfortunately, that initiative remains the domain of individual leaders rather than US commanders collectively. Authors at Fort Leavenworth, home of the US Army Mission Command Center of Excellence, observed, "the army has not fully implemented MC [mission command] because there is not uniform understanding of MC doctrine."[15] Rather than recognizing the problem as one of inadequate training and shortcomings in leaders unwilling to mentor their subordinates, the response has been one of overcomplicating an elegantly simple construct. Lengthy checklists accompany stilted prose until the belated reconsideration of mission command in 2019.[16] Mission command inventories included twenty-one "required capabilities," ten "mission command essential capabilities," and fifty-one "tasks to accomplish the required capabilities"[17] prior to the introduction of new mission command doctrine in 2019. The revision, while improved, remains a work too laden with lists.

The Australian Approach to Mission Command

Mission command is essentially about professional trust between commanders and subordinates.[18]

General Peter Cosgrove, Australian Army

There is little to distinguish the Australian Army's approach to mission command and that of Americans. Clarity of orders and intent, decentralized decision-making, and trust are the underpinnings that bring about unity of effort through the exercise of mission command in Australia's ground force as in the US Army.[19] Exercising mission command while avoiding unnecessary risk receives explicit notice in Australian joint doctrine as in that American, the objective being flexibility and adaptability the better to respond to the unexpected.[20]

Where US and Australian approaches diverge is in the amount of doctrinal guidance provided. Australian doctrine tends to better appreciate mission command's simplicity without ignoring the difficulty of its proselytization. The end sought is no different. The underlying wisdom is the same, but Australia seems satisfied that the way to propagate mission command need not require encumbering the philosophy with undue adorning that obscures rather than illuminates. Therefore, and because ultimately this is a text viewing mission command from the Australian Army's perspective, we will use its definition from here on. Restating from the chapter's opening paragraph:

> Mission command is the practice of assigning a subordinate commander a mission without specifying how the mission is to be achieved.[21]

THE ORIGINS OF MISSION COMMAND

Never tell people how to do things. Tell them what to do, and they will surprise you with their ingenuity.[22]

General George S. Patton
War As I Knew It (275 or 375)

From whence comes this approach to command? Most of those writing attribute its roots to the Napoleonic era. Having suffered at the hands of the French emperor, Prussian military leaders sought a way to replicate the Grand Armée's flexibility in battle. Napoleon's marshals understood their master's intentions and exercised the initiative necessary to act within that guidance.[23] Dispersed operations during the 1866 Austro-Prussian and 1870 Franco-Prussian Wars reinforced Prussian leaders' belief that educating junior officers in the necessity of demonstrating resourcefulness within the bounds of seniors' guidelines was the logical solution to combat's play of chance and unpredictability.[24]

7

Auftragstaktik—command based on clear but general expressions of intent and subordinates exercising freedom of judgment within those guidelines—became the norm with publication of the 1888 German field regulations.[25] Exercises in which younger officers were forced to exercise such judgment, at times even to the point of having to disobey orders, ingrained understanding that commanders' intentions took priority over specific instructions. Armor commander General Heinz Guderian recalled such an exercise of initiative in 1940 France:

> Early on the 25th of May I went to Watten to visit the *Leibstandarte* and to make sure that they were obeying the order to halt [along the Aa River]. When I arrived there I found the *Leibstandarte* engaged in crossing the Aa. On the far bank was Mont Watten, a height of only some 235 feet, but that was enough in this flat marshland to dominate the whole surrounding countryside. On the top of the hillock, among the ruins of an old castle, I found the divisional commander, Sepp Dietrich. When I asked why he was disobeying orders, he replied that the enemy on Mont Watten could "look right down the throat" of anybody on the far back of the canal. Sept Dietrich had therefore decided on the 24th of May to take it on his own initiative. The *Leibstandarte* and Infantry Regiment 'G.D.' on its left were now continuing their advance.... In view of the success that they were having I approved the decision taken by the commander on the spot and made up my mind to order the 2nd Panzer Division to move up in their support.[26]

The same authors attributing the birth of *Auftragstaktik* to defeat at the hands of Napoleon credit US and other armed forces' appreciation for its effectiveness as applied by their German adversaries during the Second World War as stimulus for eventual development of mission command in later doctrines. It is an attribution substantiated by Lieutenant General Donald Holder, one of the primary authors for both the 1982 and 1986 editions of US Army Field Manual (FM) 100-5, *Operations*:

> General Starry made the inclusion of mission command part of his initial guidance for the re-writing of FM 100-5…. One of Starry's priorities was reinvigorating our doctrinal treatment of maneuver. He generally wanted the new US *Operations* manual to parallel the current *Bundeswehr* [Army Regulation] 100/100…. I read German sources on *Auftragstaktik* and used what I learned in writing relevant portions of FM 100-5. As I recall, Huba [Wass de Czege, a fellow author for the 1982 manual] read those documents too. We were both influenced by the Howard-Paret translation of Clausewitz's *On War*, Rommel's *Infantry Attacks*, Manstein's *Lost Victories*, and by the post-WWII interviews with Wehrmacht leaders…. I think that it's important to note that the US Army's knowledge and practice of mission command went back a lot further than 1982. Armor branch inculcated the method into its youngest officers and at field grade level those officers would often argue for that method of command in brigade and division operations. There was conscious disagreement between advocates of directive command and mission command and a general tendency for infantry officers to prefer the more restrictive form. (I thought that preference came from their training in airborne, air assault and night ops, which all legitimately require closer control.) Since that was the case, the domination of maneuver advocates at TRADOC headquarters (Starry, Otis, Richardson) had a big effect on the direction of Army doctrine. They succeeded in re-emphasizing mission command/*Auftragstaktik* in our doctrine even though the debate about the proper balance between directive command and mission command continued in the field.[27]

Post-World War II re-adoption of mission command (or perhaps more accurately, re-recognition of its value) was slow in coming. The United States experienced years of detailed direction by commanders before those 1982 and 1986 *Operations* manuals formally suggested "decentralization demands subordinates who are willing and able to take risks and superiors who nurture that willingness and ability in their subordinates."[28]

US Army rediscovery or otherwise notwithstanding, effective commanders have exercised the principles underlying mission command for millennia. The Roman commander Vespasian chose his son Titus to complete the empire's suppression of a 1st-century AD uprising in Galilee, Samaria, and Judea after the father became emperor in the year 69. Titus had demonstrated his ability while campaigning alongside Vespasian in the preceding years. He was therefore trusted to complete the campaign following the tumultuous year of the four emperors as 69 would later come to be known. Captain John Pershing would similarly demonstrate understanding of his political if not military masters' intentions when he resisted immediate seniors' demands that he combat Moros in the southern Philippines in the aftermath of the Spanish-American War, instead judging that the US "can well afford to wait and exhaust every effort to establish friendly relations." When he did resort to combat, he made it clear that only specific clans rather than Moros in general were his adversaries.[29] The result was success where others had failed and, three years later, Pershing's promotion from captain to brigadier general, a leap over 862 seniors in rank.[30] Exercised in 1903, included in the US Army's 1939 field regulations, revived in 1980s doctrine, and formally given the moniker "mission command" in 2003, the concept is one likely as old as the first enlightened military leader who found it necessary to send a portion of his force over a ridgeline or along a separate route in preparation for battle.

FUNDAMENTALS UNDERPINNING MISSION COMMAND

Vespasian's choice of son Titus to assume command of the campaign in Judea was founded on far firmer stuff than nepotism alone. The emperor knew of his choice's *expertise* as a commander and *experience* relevant to the tasks he would have to perform. Titus had earlier demonstrated both—and his *reliability*—when commanding away from his father's direct oversight. Together these and other factors meant Vespasian *trusted* Titus. Trust must obviously underlie decentralization, trust in subordinates' judgment and, in turn, subordinates' trust that their commander will back their decisions should those decisions have been made in faith with seniors' intentions. *Familiarity*, obviously a part of the Vespasian-Titus relationship, will also play a significant role in determining the extent of operational freedom granted. That scope will differ from individual to individual. The well-known junior commander with demonstrated ability to function without close supervision merits less oversight than one less familiar or proven. The receiving commander should provide closer supervision, grant less freedom of action, and give more specific guidance when dealing with what are from his perspective unproven leaders. With such greater control a commander acknowledges his own ignorance: the less familiar he is with subordinates' capabilities, the greater the need for him to ensure his guidance is appropriate to the resource provided. Time before pending operations and nature of the mission will influence the scope of leeway bestowed, time as it may reassure the commander regarding these new subordinates' abilities, mission because the most brilliant leader in some situations might require increased supervision when pursuing objectives of another type. World War II German General Friedrich-Wilhelm von Mellenthin drew on his considerable experience when noting "commanders and subordinates start to understand each other during war. The better they know each other, the shorter and less detailed the orders can be."[31] His words ring true regardless of the type of operation at hand.

While the mission statement may be the same for all, the level of detail in instructions to each commander should reflect the degree to which the senior leader authorizes decentralized decision-making. Greater familiarity and trust combined with a high level of subordinate expertise would tend to result in lesser risk associated with decentralization. Granting the same to a less proven or known individual would qualify as imprudence.

How to cultivate effective mission command? Via training and command responsibility. Training in military schoolhouses where junior noncommissioned and commissioned officers learn their trade, where mid-grade leaders learn staff and command tradecraft, and seniors ready for the pinnacles of responsibility. Training in units, where exercises force decision-makers to deal with the unexpected and allow senior commanders to demonstrate that well intentioned if less-than-perfect judgments are not only allowable but demanded. Training via self-education guided by mentors that ensures subordinates read Grant, Slim, and others whose command styles demonstrate mission command at its best. And training through one-on-one evaluations in which the overly conservative and risk-averse learn that his or hers is not an acceptable form of leadership. Trust, familiarity, and expertise gained in training provide foundation stones for mission command's application during operations.

MISSION COMMAND DURING 21ST-CENTURY OPERATIONS

Our entrance into Kandahar and Baghdad marked the beginning of a transition to decentralization and empowerment for our army upon which we continue to build. Our collective experience with mission command has evolved over the past decade of conflict, and mission

command has emerged as one of the central tenets underpinning how
our army currently fights.[32]

Lieutenant General (US Army) David G. Perkins

Our discussion to this point makes it clear that mission command should be conditional rather than absolute in application. One size does not fit all. A commander fortunate enough to have key subordinates with whom he has long worked, trusts, and who have proven themselves competent in operations like those pending will require less direction and supervision than individuals less familiar, not as trusted for whatever reason, or who lack the experience to merit greater freedom of action. The task at hand, nature of the threat, environment, and other factors will likewise influence the character of mission command exercised. We have also noted that even familiar, completely trusted, and very experienced subordinates require more command guidance under some circumstances than during others. Resource availability will also influence the extent of decentralization granted. Freedom of action with regard to employing one's own forces will logically be greater than in allocating low-density assets such as air or artillery fire support.[33] These observations apply to members of one's own service, other national assets, and during contingencies involving a multinational, whole of government, or comprehensive approach (e.g., those incorporating nongovernmental, inter-governmental, or commercial partners).

Comfort in exercising mission command is likewise a matter of military culture. Its US resurrection during the last decade of the Cold War was in part a response to perceptions that fighting a more numerous Warsaw Pact foe on Western Europe's compartmented terrain where communications might fail meant leaders would be unable to personally direct all of their command elements. The agility inherent in mission command practice was also seen as an advantage over those opponents, adversaries for whom extensive variation from plans was antithetical.[34] Yet there were considerable variations in command approaches even within North Atlantic Treaty Organization and other national militaries considered aligned with the United States.[35] The Israel Defense Forces, thought to favor highly decentralized tactical operations, proved uncomfortable with the full extent of decentralization associated with *Auftragstaktik*. Its leaders instead opted for what two authors labeled "selective control" in which those exercising higher-echelon oversight provided mission-type orders and expected initiative even as they tracked operations in great detail, remaining ever prepared to intervene should a situation appear to be beyond a subordinate's capabilities or opportunity arise that otherwise might be lost.[36] Israeli control has apparently become further centralized in succeeding years. While ground force units were assigned increased numbers of air support liaison personnel during 2014 Operation Protective Edge in Gaza, those at the sharp end had to request clearance for danger close strikes from a centralized authority remote from the battlefield.[37] Some contrast British command approaches (and presumably those of the Australian, Canadian, and New Zealand militaries with whom they share cultural and historical ties) with those American, the former relying on assigned objectives communicated in quite general terms while US leaders provide more detailed guidance in their orders. This greater specificity is thought to dictate more in the way of how objectives are to be accomplished, resulting in less freedom of action by commanders on the receiving end.[38]

Variations in approach are not limited to those between national militaries. Other-than-armed forces organizations have in recent years recognized value in adopting a mission command-type philosophy. The Australian Fire and Emergency Services

Council (AFAC) finds the approach beneficial during the conduct of its often geographically-dispersed operations. Similar to military conceptions of mission command, AFAC leaders are to communicate a commander's intent and ensure subordinates receive the resources necessary to succeed in serving both mission-specified ends and those implied by that intent.[39]

EXERCISING MISSION COMMAND

Divisions…under my command…fought on a front of seven hundred miles, in four groups, separated by great distances, with no lateral communications between them and beyond tactical support of one another…. Commanders at all levels had to act more on their own; they were given greater latitude to work out their own plans to achieve what they knew was the Army Commander's intention. In time they developed to a marked degree a flexibility of mind and a firmness of decision that enabled them to act swiftly to take advantage of sudden information of changing circumstances without reference to their superiors…. This acting without orders, in anticipation of orders, or without waiting for approval, yet always within the overall intention, must become second nature in any form of warfare.[40]

Field Marshal (British Army) William Slim
Defeat Into Victory

Subordinates experience and expertise, their demonstrated ability to exercise good judgment under relevant operational conditions, a commander's familiarity with those individuals, the extent of trust that senior leader imbues given these and other considerations: all are factors influencing the nature of guidance given to and freedom of action bestowed on each of those subordinates by a commander. The subordinate's responsibilities within the context of mission command are thus far less clear in our discussion above. Clearly there must be understanding of why one individual receives more detailed guidance and closer supervision than another. Trust will play a part, but trust has many components. Lesser trust by no means need imply a senior questions the judgment or reliability of a junior, but rather that those are qualities as of yet unmeasured. Trust—from above to below and vice versa—comes only with demonstrated performance, validation, and the passage of time. Even the most dependable subordinate will occasionally find the diligent commander ensuring his or her actions fall within bounds of the senior's intent. Subordinates' have a responsibility both to operate within those bounds and educate their senior commanders when they lead a unit less familiar to those above them in the chain of command.

This requirement to educate assumes a maturity in subordinate leaders that a commander might well find absent in some juniors, certainly in those new to their military careers. That mission command has proved so elusive for some in the US military despite its long being promoted demonstrates the need for more effective training of both those senior and junior leaders. Mission command is elegantly simple in construct but arduous in application. Only with effective training can a force hope to harvest its considerable benefits.

TWELVE PERSPECTIVES ON THE AUSTRALIAN ARMY APPROACH TO MISSION COMMAND

The lesson for me was that despite all the thought and planning that can go into preparing for and conducting a mission, there is always a bigger picture that may not be readily evident.[41]

Lieutenant General (Australian Army) John Caligari
"Operation SOLACE (Somalia 1993) and the lessons learned"

Eleven Australians and one Canadian analyze applications of Australian mission command in the chapters to follow. Dr. Peter Pedersen, World War I historian and former commander of the 5th/7th Battalion, the Royal Australian Regiment (RAR), turns his attentions to that conflict while Dr. Peter Dean draws on his longtime study of the Second World War in looking at Australian-US command relationships in the Pacific theater. Canadian Dr. Meghan Fitzpatrick brings her considerable knowledge of the Korean War to view command relationships during the last instance of the Australian Army subordinating a unit of brigade size to the British Army. Dr. Bob Hall, infantry platoon commander with 8RAR in Vietnam, considers command relationships characterized by a range of emotions that include frustration, befuddlement, and respect depending on the personalities at hand. Lieutenant General John Caligari was operations officer for 1RAR as a major in Somalia. His were experiences during humanitarian operations in a non-permissive environment. Dr. John Blaxland served as brigade intelligence officer in 1999 East Timor. Lieutenant General John Frewen views command relationships during the Regional Assistance Mission to Solomon Islands in 2003 during which he led the some 1,800-strong, five-nation military contingent supporting that undertaking. Major General Tony Rawlins draws on first-hand experiences from his 2006-2007 tenure as commander of Overwatch Battle Group West-Two in Iraq, as do Brigadier Chris Smith and Brigadier Ian Langford from service in Afghanistan, the latter providing a special operations perspective. Major General Chris Field offers the too often overlooked but crucial viewpoint of a military officer who supported domestic disaster relief operations, in his case those during and after the devastating 2010-2011 Queensland floods. Major General Roger Noble concludes the book, pushing back from the table to consider Australia's approach to mission command in light of his recent leadership of the Australian Army's 3rd Brigade. Together these considerations present readers an opportunity to appreciate a highly professional army's approach to mission command across a broad range of challenges.

CONCLUDING OBSERVATIONS

Mission command must be endorsed and practiced at all levels in order to be effective. This requires implicit trust between and across all elements of the land force, with junior leaders possessing a detailed understanding not only of the immediate tactical commander's intent, but also of the broader operational and strategic situation. The subordinate is then expected to apply individual judgment in achieving the commander's intent, regardless of changing situations…. Army must actively create the climate and foster behavior that produces a mission command culture. [42]

Australian Army Land Warfare Doctrine 1
The Fundamentals of Land Power

The discussion above establishes the conditional nature of a commander's applying mission command in light of subordinates' abilities. What should be unconditional, however, is the approach's application throughout an armed forces. Having only select commanders adhere to its tenants is similar to developing a professional police force without addressing the remainder of a legal system: the police arrest perpetrators only to find corrupt judges release the recalcitrants or prisons free them in return for bribes. Historically, times of relative peace in particular see less confident or able leaders practicing risk aversion. Fear of a subordinate making a mistake that might threaten a senior leader's career tightens centralization. [43] Enhanced communications technologies become implements of intrusion on junior leader decision-making. Those in helicopters overhead at least realized that jungle foliage or elephant grass blocked much of their vision in Vietnam. There are no such obvious filters when looking at a computer screen's false clarity. "Train to trust" and "train to take appropriate risk" must be building blocks for propagating mission command. The commander who tolerates otherwise is an obstacle to that nurturing. Commending rather than condemning wisely taken decisions that result in undesirable outcomes is a necessary yet too rare event. We noted that late 19th and early 20th-century Prussian and German military exercises deliberately forced subordinates to vary from mission dictates within the constraints of their commander's intent. The US Army instead calls for mission command strategies, systems, and checklists.[44] No list can account for every possible scenario; that with one hundred items helps but little when reality presents situation 101. Checklists undeniably have their place. Failing to account for one item in preflight preparations invites catastrophe; faulty pre-jump checks can send a paratrooper to his death. Military operations are more akin to navigating a kayak in a fast-flowing river than preparing for aircraft takeoff and are thus less amenable to mechanical practices.

Operations in these 21st-century opening years increasingly demand a comprehensive approach—one involving all services, multiple nations with several government agencies from each, and capabilities only other-than-government organizations such as NGOs, inter-governmental, and commercial enterprises can bring to the table. Decentralization is a given; such operations will never see unity of command. Unity of effort is the perhaps achievable goal with various organizations' efforts orchestrated via a commonly agreed upon intent. Mission command's underlying foundation stones—a clear intent, trust, initiative, understanding of context and objectives sought, familiarity with subordinates, decentralization, and the courage to accept risk—are graspable regardless of background. Leaders, military and civilian alike, recognize the need for better conducting comprehensive approaches to campaigns. That approach, like mission command, remains an unfulfilled goal. Mission command offers a means of achieving the orchestration essential to a successful comprehensive approach.

The insights provided by the chapters to follow reinforce the value of mission command. They also warn of the consequences inherent in failing to practice it effectively. Continued enhancement of communications technologies will be a tool for undue centralization in the hands of leaders unversed in—or unwilling to apply—mission command. Increasing reliance on such technologies should reinforce calls for better inculcating mission command throughout a military. It will be to their commander's intent that subordinates will have to turn when those communications fail due either to enemy antipathy or nature's hand. Organizations unable to practice effective mission command will find themselves at a disadvantage when facing commanders who "receive general operation guidelines but have significant autonomy to run their own operations" as do those in the Islamic State of Iraq and Syria (ISIS).[45] The remaining chapters offer an opportunity to draw on the experiences of an able ally to aid in realizing the elusive goal of its effective application. Those experiences reveal that the challenges inherent in mission command include not only persuading over-controlling leaders to adapt their ways but also convincing leaders and subordinates alike that, properly applied, mission command reinforces rather than replaces the age-old dictum that soldiers do well what leaders check.

ENDNOTES

1   Ulysses Simpson Grant, *Personal Memoirs of General Ulysses S. Grant* (NY: Cosimo, 2007), 278.

2   Land Warfare Doctrine 1, *The Fundamentals of Land Power* (Canberra: Australian Army, 2014), 45.

3   US Army Field Manual 100-5, *Tentative Field Service Regulations, Operations* (Washington, D.C.: US Government Printing Office, 1939), 60.

4   John Case, "The Exigency for Mission Command: A Comparison of World War II Command Cultures," *Small Wars Journal* (November 4, 2014), www.smallwarsjournal.com/printpdf/17005 (accessed July 7, 2015).

5   Grant, *Personal Memoirs*, 278.

6   *DOD Dictionary of Military and Associated Terms*, Washington, D.C.: US Joint Chiefs of Staff (June 2020), 144.

7   Army Doctrine Reference Publication (ADRP) 6-0, *Mission Command* (Washington, D.C.: Headquarters, Department of the Army, May 17, 2012), 1-1. "Mission orders" are in turn defined on page Glossary-3 of the same publication as "directives that emphasize to subordinates the results to be attained, not how they are to achieve them."

8   Ibid, 2-4.

9   Army Doctrine Publication (ADP) 6-0, Mission Command: Command and Control of Army Forces (Washington, D.C.: Headquarters, Department of the Army, July 2019), Glossary-3.

10 The United States Navy does not employ mission command per se. It does, however, have among its command approaches "command by negation" that shares a number of characteristics with mission command.

11 *Mission Command*, 2-3.

12 Russell W. Glenn, "The Commander's Intent: Keep It Short," *Military Review* 67 (August 1987): 51.

13 US Army Field Manual 100-5, *Operations* (Washington, D.C.: Headquarters Department of the Army, August 20, 1982), 2-2.

14 Eitan Shamir, "The Long and Winding Road: The US Army Managerial Approach to Command and the Adoption of Mission Command (Auftragstaktik)," *Journal of Strategic Studies* 33 (October 2010), 663.

15 Mission Command Center of Excellence, "U.S. Army Mission Command Strategy FY 13-19," Fort Leavenworth, KS: US Army Combined Arms Center, 2013, http://usarmy.vo.llnwd. net/e2/c/downloads/312724.pdf (accessed July 9, 2015), 2.

16 The objective of mission command, for example, was articulated as "unity of effort to effectively integrate and synchronize operational and institutional forces' roles and responsibilities to implement MC across the doctrine, organizational structures, training, materiel, leadership and education, personnel, and facilities (DOTMLPF) domains." Army Doctrine Reference Publication (ADRP) 6-0, *Mission Command* (Washington, D.C.: Headquarters, Department of the Army, May 17, 2012), 1-1; and "U.S. Army Mission Command Strategy FY 13-19," 1. DOTMLPF is an acronym for doctrine, organization, training, materiel, leadership and

education, personnel, and facilities. Fortunately the 2019 *Mission Command* doctrinal manual stepped away from these cumbersome approaches.

17 "U.S. Army Mission Command Strategy FY 13-19," MC-REQUIREMENTS-2.

18 Australian Defence Doctrine Publication (ADDP) 00.1, *Command and Control* (Canberra: Department of Defence, May 27, 2009), 2-8.

19 Australian Defence Doctrine Publication (ADDP) 3.0, *Campaigns and Operations* (Canberra: Department of Defence, July 12, 2012), 2-3.

20 *Command and Control*, 2-11.

21 Australian Army, *The Fundamentals of Land Power, Land Warfare Doctrine* (LWD) 1 (Canberra, ACT: Australian Army, 2014), 45.

22 George S. Patton, *War as I Knew It* (Boston: Houghton Mifflin Harcourt, 1995), 357, as cited in Douglas A. Pryer, "Growing Leaders Who Practice Mission command and Win the Peace," *Military Review* (November-December 2013): 32.

23 John Case, "The Exigency for Mission Command: A Comparison of World War II Command Cultures," *Small Wars Journal* (November 4, 2014), www.smallwarsjournal.com/printpdf/17005 (accessed July 7, 2015). Prussian leaders' acceptance of so innovative an approach was far from unanimous. Debate continued throughout the 19th and into the 20th centuries.

24 Eitan Shamir, *Transforming Command: The Pursuit of Mission*

Command in the U.S., British, and Israeli Armies (Stanford, CA: Stanford University Press, 2011), 37.

25 Douglas A. Pryer, "Growing Leaders Who Practice Mission command and Win the Peace," *Military Review* (November-December 2013): 31.

26 Heinz Guderian, *Panzer Leader,* Trans. Constantine Fitzgibbon (London: Futura, 1977), 117. Antulio J. Echevarria uses this example very effectively in his article "Auftragstaktik: In Its Proper Perspective," *Military Review* 66 (October 1986): 50-56.

27 L. Donald Holder (Lieutenant General, US Army, retired) email to Dr. Russell W. Glenn, Subject: Re: Origins of a "mission command" type philosophy in 1982 and 1986 FM 100-5s, August 1, 2015. "Starry, Otis, Richardson" refers to US Army Training and Doctrine Command commanders Donn A. Starry, Glenn K. Otis, and William R. Richardson, respectively, who served consecutively from July 1, 1977-June 30, 1986.

28 Antulio J. Echevarria, "Auftragstaktik: In Its Proper Perspective," *Military Review* 66 (October 1986): 56.

29 Brian McAllister Linn, *Guardians of Empire: The U.S. Army and the Pacific, 1902-1940* (Chapel Hill, NC: The University of North Carolina Press, 1997), 38.

30 Gene Gurney, "Foreword," in John J. Pershing, *My Experiences in the World War* (Blue Ridge Summit, PA: Tab Books, 1989), xv.

31 As quoted in Eitan Shamir, *Transforming Command: The Pursuit of Mission Command in the U.S., British, and Israeli Armies* (Stanford, CA: Stanford University Press, 2011), 106.

32 David G. Perkins, "Mission Command," *Army* 62 (June 2012): 31.

33 Ministry of Defence of the Netherlands, Joint Doctrine Publication 5: *Command and Control*, (Doctrine Branch, Netherlands Defence Staff, March 16, 2012), 59.

34 Douglas A. Pryer, "Growing Leaders Who Practice Mission command and Win the Peace," *Military Review* (November-December 2013): 32.

35 As would be expected, these variations continue to exist today. The Dutch Army's March 2012 Command and Control joint publication lists three pillars of command: "the network approach, mission command, and the effects-based approach." Though Netherlands leaders felt effects-based-type operations (EBO) served them well in southern Afghanistan, EBO were removed from US joint doctrine in late 2008 when US Joint Forces Command commander James N. Mattis concluded the approach had "been misapplied and overextended to the point that it actually hinders rather than helps joint operations." James N. Mattis, "USJFCOM Commander's Guidance for Effects-based Operations," *Parameters* 38 (Autumn 2008): 18.

36 David S. Alberts and Richard E. Hayes, "Command Arrangements for Peace Operations," paper in the Command and Control Research Program (CCRP) publication series, 1995, 69, http://www.dodccrp.org/events/12th_ICCRTS/CD/library/html/pdf/Alberts_Arrangements.pdf_(accessed July 27, 2015). Alberts and Hayes' work is summarized in Keith Stewart, "Mission Command: Problem Bounding or Problem Solving?" *Canadian Military Journal*, undated, http://www.journal.forces.gc.ca/vo9/no4/09-stewart-eng.asp (accessed July 8, 2015).

37 Russell W. Glenn, *Short War in a Perpetual Conflict: Implications of Israel's 2014 Operation Protective Edge for the Australian Army* (Canberra, Australia: Australian Army, 2016).

38 David S. Alberts and Richard E. Hayes, "Command Arrangements for Peace Operations," paper in the Command and Control Research Program (CCRP) publication series, 1995, 70, http://www.dodccrp.org/events/12th_ICCRTS/CD/library/html/pdf/Alberts_Arrangements.pdf (accessed July 27, 2015).

39 Euan Ferguson, "Mission Command for Fire and Emergency Managers: A Discussion Paper," Australian Fire and Emergency Services Council (AFAC), May 2014, http://www.cfabellarine.com/uploads/1/3/0/0/13001256/mission_command_discussion_paper_may_2014.pdf (accessed July 7, 2015).

40 William Slim, *Defeat Into Victory* (London: The Reprint Society, 1957), 525.

41 John Caligari, "Operation SOLACE (Somalia 1993) and the lessons learned," *Australian Defence Force Journal*, no. 184 (2011): 17.

42 Land Warfare Doctrine 1, *The Fundamentals of Land Power* (Canberra: Australian Army, 2014), 45.

43 Gary Luck, "Mission Command and Cross-Domain Synergy," Insights and Best Practices Focus Paper, Suffolk, VA: Deployable Training Division, Joint Staff J7, March 2013, 4.

44 Eitan Shamir, "The Long and Winding Road: The US Army Managerial Approach to Command and the Adoption of Mission Command (Auftragstaktik)," *Journal of Strategic Studies* 33 (October 2010), 666.

45 Eric Schmitt, and Ben Hubbard, "ISIS Leader Takes Steps to Ensure Group's Survival," *The New York Times* (July 20, 2015), http://www.nytimes.com/2015/07/21/world/middleeast/isis-strategies-include-lines-of-succession-and-deadly-ring-tones.html?_r=0 (accessed July 22, 2015).

2

MISSION COMMAND AND THE AUSTRALIAN IMPERIAL FORCE

Dr. Peter Pedersen

Throughout the First World War, Australian formations fought as part of British armies and, for much of the war, were led by British officers. Though the term "mission command" had yet to enter the English military lexicon, both the Australian and British commanders were acquainted with the principles it entailed. First published in 1905 and based on the lessons learned from the war in South Africa (1899–1902), the British Army's *Field Service Regulations (FSR)* stated that "decentralisation of command and a full recognition of the responsibilities of subordinates in action" were "absolutely necessary." *FSR 1909* set the principle out in a form that has a contemporary ring: "The object to be attained, with such information as affects its attainment, should be briefly but clearly stated while the method of attaining the object should be left to the utmost extent possible to the recipient, with due regard to his personal characteristics."[1] The various arms and services manuals reflected *FSR's* principles.[2] For example, *Infantry Training 1914* stated, "It is essential that superior officers, including battalion commanders, should never trespass" on their subordinates' sphere of action.[3]

PEACE TO WAR

As a dominion of the British Empire, Australia had increasingly cooperated with Britain in imperial defense matters before the war. *FSR* inevitably reached into the Australian Army as standardization in military organization, equipment, and procedures was part of the process. In an award-winning essay on the lessons for the Australian Army of the American Civil War Wilderness campaign of 1864, one of Australia's leading citizen soldiers, Colonel John Monash, almost echoed its words: "The subordinate should not be hampered by too precise or detailed instructions, but is to be encouraged to act according to local circumstances."[4] In *100 hints for company commanders*, a pamphlet that he wrote for junior leaders while commanding the 13th Infantry Brigade in 1913-14, Monash directed that they must "never interfere with the performance of a duty for which a subordinate is responsible, unless his performance of it is incorrect."[5] Here Monash also acknowledged that circumstances could necessitate a higher commander's intervention. In other words, mission command depended on finding the happy medium between control and a free hand.

Yet the practical application of mission command in the pre-WWI Australian Army was impossible. Lacking formed regular units, it was a part-time citizen force in which service became compulsory in 1911. Sixteen days were allocated for annual training, only eight of them in camp where a few senior officers might get the chance to handle a brigade for a day or two.[6] This was hardly conducive to developing the skills of both leaders and led – or the trust between them – that are essential for a flexible approach to command in battle. That Monash found it necessary to write the orders for one of his battalion commanders at the 13th Brigade's camp at Lilydale in February 1914 was not surprising.[7]

As Australia's Defence Act restricted the army to home defense, a voluntary force was raised for service abroad when war came. Initially comprising four infantry brigades (the first three forming the 1st Australian Division) and three light horse brigades, the Australian Imperial Force (AIF) reflected the army's shortcomings. Australia's senior soldier, Major General William Bridges, led both the AIF and the 1st Division. He had served mainly in staff and instructional postings, seeing little time in command even at the regimental level. Monash had fleetingly commanded a brigade. The other three infantry brigadiers lacked even that experience. Practical knowhow was limited across the wider leadership. Most of the 1st Division's 631 officers came from the citizen forces.[8] The rush to raise the AIF precluded addressing these command deficiencies until its arrival in Egypt at the end of 1914. Once there, the 1st Division and the composite New Zealand and Australian (NZ and A) Division, which contained most of the other Australian formations, were grouped in the Australian and New Zealand Army Corps (ANZAC). By mid-February 1915, the 1st Division had begun brigade maneuvers, which did not test Bridges as divisional commander, and the NZ and A Division, having rushed lower-level training, was just starting divisional ones. But time was short.

GALLIPOLI

Six weeks later both divisions had left Egypt as part of the Mediterranean Expeditionary Force (MEF) formed under General Sir Ian Hamilton. The MEF was to seize Turkey's Gallipoli peninsula to allow the fleet to pass uncontested through the Dardanelles and reach Constantinople, hopefully prompting Turkey's surrender. Taking the peninsula required the MEF to carry out what is among the most complex of military operations—an opposed amphibious landing—with little training and in the era before real-time communications. Hamilton did not let the difficulties affect his "obsession with the freedom of his subordinates to run their own show."[9] He gave Lieutenant General Sir William Birdwood, the ANZAC's British commander, a free hand apart from instructing him to make a subsidiary landing near Gaba Tepe promontory. Success would cut the Turkish lines of communication to Cape Helles, twenty kilometers south, where the British would make the main landing. Whether or not the first ANZAC brigade ashore would advance beyond the line that ran up the second of three ridges to Chunuk Bair, the key height on the dominating Sari Bair range, "must be left to your discretion," Birdwood was told.[10]

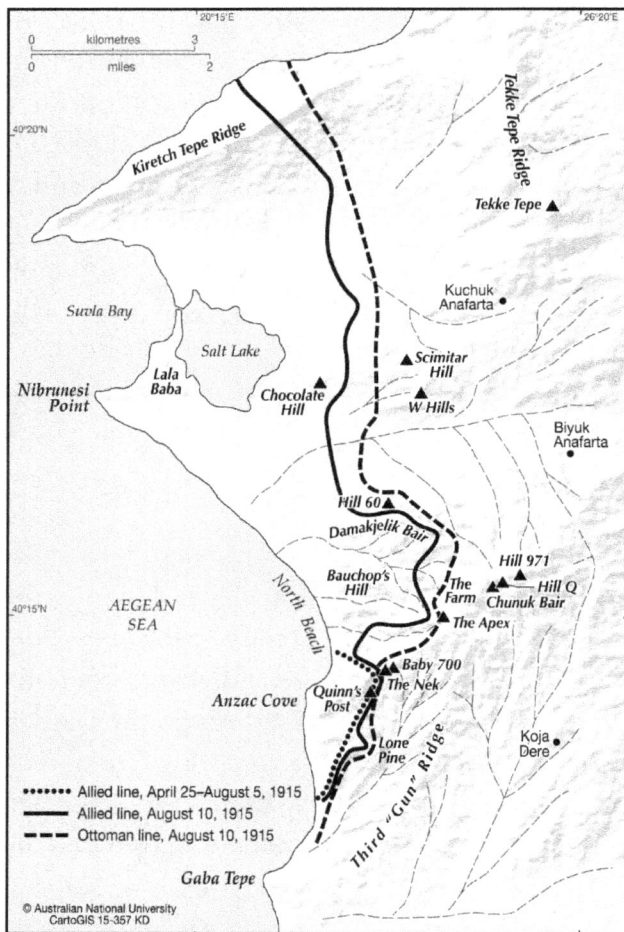

Map 2-1: Anzac and the area of the August offensive[11]

Figure 2-1: Troops of the 2nd Brigade landing at Anzac Cove early on the morning of April 25, 1915.[12]

Birdwood decided that the line should include Hills Q and 971, the other two peaks of the Sari Bair range, and the third ridge inland so that guns enfilading the landing beach from Gaba Tepe, where the ridge ended, could be quickly captured. The resulting bridgehead would be much larger than Hamilton had envisaged. Birdwood gave Bridges the task of establishing it. He in turn tasked Colonel Ewen Sinclair-Maclagan and his 3rd Brigade of Australians to lead the assault.[13] Though Sinclair-Maclagan was a decorated British regular and Bridges' trusted friend, he had never commanded above company level and was also pessimistic. He gloomily told Bridges that Gaba Tepe would be "almost impregnable for my fellows" were it strongly held and that if the ridges beyond were also well defended his brigade would never be seen again.[14] Sinclair-Maclagan's attitude would prove not conducive to the flexibility needed to deal with the unexpected difficulties he would confront as the senior commander ashore during the landing's first few hours.

Landing at 4:30 am on April 25, 1915, the 3rd Brigade found itself over a kilometer farther north than planned. Though badly intermixed on the beaches of what history would label Anzac Cove, the soldiers quickly scaled First Ridge and paused to reorganize. The situation was not irretrievable. They were as close to Third Ridge as would have been the case barring the initial navigation error and much further away from the guns at Gaba Tepe, from which Anzac Cove was largely sheltered anyway.[15] Opposition was almost non-existent.

Yet Sinclair-Maclagan was convinced that reaching his objective of Third Ridge was no longer feasible. He halted his brigade on Second Ridge, one and a half kilometers short of the assigned goal. The 2nd Brigade under Colonel James McCay was landing at this time. Sinclair-Maclagan urged McCay to strengthen the right flank, which faced Gaba Tepe.[16] McCay acquiesced, believing that Sinclair-MacLagan had effectively assessed the situation from his forward position. He thereby jettisoned his task of capturing the vital Sari Bair heights on the left. Sinclair-McLagan's timidity had administered last rites to the 1st Division's mission.[17]

No resurrection was forthcoming from Bridges. The division commander had landed at 7:20 am. Over 8,000 Australians were ashore by then with fewer than 500 Turks directly facing them.[18] Neither Bridges nor his chief of staff, Colonel Brudenell White, could see anything to prevent an advance to Third Ridge; neither did anything to revive the assault.[19] After the mid-morning arrival of Turkish reserves, the battle swayed back and forth on Second Ridge. The Australian line had stabilized by nightfall, but Bridges knew little of the situation other than that casualties were heavy. He recommended evacuation to Birdwood, who had come ashore briefly at 3 pm, later re-embarking "without serious misgiving."[20] Birdwood passed the decision to Hamilton whose hands-off approach at the earlier Helles landing, even when he saw that it was going seriously amiss, almost resulted in a catastrophe. Ironically,

Hamilton now made his solitary intervention of the day, telling Birdwood "to dig, dig, dig, until you are safe." By day's end the ANZAC clung to an area of but two square kilometers instead of reaching across the peninsula as Hamilton had wanted. That area became known as Anzac.

Troops at the sharp end had fought a host of minor isolated actions that proved harsh introductions to what would be known as mission command today for junior commanders. The landing had also demonstrated that mission command at higher echelons, so easy to preach in peacetime, was far less simple to achieve in combat. Meeting Hamilton's intentions first rested on Sinclair-MacLagan making the right decisions. The 3rd Brigade's commander failed to do so. Bridges and Birdwood "made no decisions at all," compounding their subordinate's miscarriage of judgment.[21] Neither applied the required control. There had been no striking of the happy medium between the controlling and free hand.

Finding that balance proved to be more hit or miss than conscious process as the campaign progressed. Senior commanders allowed subordinates to occupy sectors largely as the latter saw fit. Major General Alexander Godley, British commander of the NZ and A Division to which the 4th Australian Brigade under Monash belonged, simply told Monash that he "*must* ensure that the *head* of your ravine is held and strongly entrenched."[22] Godley was referring to Quinn's Post at the apex of the Anzac line, a position whose loss would jeopardize the entire Anzac position. The Turks would later break into Quinn's Post on May 29, 1915. Colonel Harry Chauvel, the Australian light horse commander who had subsequently assumed responsibility for the sector, overruled Major Harry Quinn after whom the post was named and who knew it intimately. Quinn urged an outflanking move to clear the Turks but Chauvel, telling Quinn how to carry out the mission, insisted on a charge over the top.[23] The charge succeeded but Quinn was killed.

In August the Anzac area became the focus of Hamilton's main effort with the launching of an offensive to seize the Sari Bair heights. Columns would make a northerly breakout from Anzac and attempt to scale the almost unclimbable—and hence largely undefended—coastal approaches to them. Birdwood intended that the 1st Division's opening attack on Lone Pine, on Anzac's southern flank, during the afternoon of August 6th, would distract the Turks before the breakout a few hours later. Both the British Army's Major General Harold Walker, commander of the 1st Division since Bridge's death in May, and Brudenell White opposed the idea, fearing an overwhelming Turkish response. That was precisely the effect Birdwood sought. He overruled them, and Lone Pine fell to the 1st Division in ferocious fighting.[24] Early on August 7th, the 3rd Light Horse Brigade struck the Nek on Anzac's northern flank to distract the Turks from the New Zealand advance on Chunuk Bair. Elderly and ill, Brigadier General Fred Hughes allowed his brigade major, Lieutenant Colonel Jack Antill, to control the operation. The Turks annihilated the first line of light horsemen. Antill dismissed protests that to continue amounted to murder, sending three more lines to their deaths. Though rattled, Hughes did nothing to counter Antill's ill-advised orders.[25]

Monash's column was to take Hill 971, but he found his ability to control its advance hamstrung by the instructions of his superior, Brigadier General Vaughan Cox, to remain at the center of the brigade instead of near its head.[26] Cox commanded the 29th Indian Brigade, which was following Monash's brigade. Unfamiliar with Monash, Cox apparently wanted him, as the lead formation's commander, readily at hand. Monash consequently had to go forward to get his brigade moving again after it had stalled on becoming lost. Badly weakened by dysentery, it eventually halted well short of Hill 971. In its next attempt on Hill 971, Monash made no effort to control the attack at all, remaining in his headquarters after putting his senior and most trusted battalion commander in charge. He failed to move forward even when he knew his brigade was in severe trouble.[27] Exhaustion is the only explanation for Monash's inaction—he left no record himself. The 4th Brigade was virtually destroyed. Both the August offensive and a concurrent landing at Suvla Bay ended in failure. The Allies evacuated the Gallipoli peninsula at the end of 1915.

EXPANSION

The AIF returned to Egypt, where the 1st Division and the 2nd Division, which had reached Anzac as the campaign wound down, were joined by the 4th and 5th. The latter two had been formed largely by splitting veteran battalions and bringing the resulting half-battalions up to strength with replacements. Raised in Australia, the 3rd Australian Division had sailed directly to England. These additional formations, and the creation of a New Zealand division, necessitated replacement of the ANZAC with two new formations, I ANZAC under Birdwood and II ANZAC under Godley. Birdwood's view that no Australians were capable of leading the newly formed infantry divisions riled the Australian government. A compromise saw Cox get the 4th Division and McCay, an Australian, the 5th.[28] Major General George Legge, architect of Australia's pre-war compulsory training scheme, remained in command of the 2nd Division. Monash was given the 3rd Division in mid-1916. Commanders of the new brigades and battalions came from within the AIF. Most had proven themselves at Anzac.[29] Though they lacked experience in their new positions, a leavening of operational knowhow now existed at all levels. I and II ANZAC left for the Western Front in March and June 1916 respectively. The new Anzac Mounted Division, led by Chauvel, remained in Egypt.

THE WESTERN FRONT

Given its status as the war's main theater, it is unsurprising that the Western Front was where most of the AIF served. It was an environment dominated by trenches, machineguns, artillery, and barbed wire that gave defense the upper hand. An open flank had offered a way around defenders at Gallipoli but the attempt to capitalize on it in August 1915 failed. There was no such way around on the Western Front. Trenches stretched from the North Sea to the Swiss border and could only be attacked frontally. Attacks were backed by the full weight of modern military technology in the form of massed artillery, aircraft, and gas. Yet influencing attacks after initiation was all but impossible given the inability to obtain real-time information on progress. Attacks were therefore highly structured affairs in which planners attempted to account for every eventuality. Such conditions were new to the AIF's commanders. So was the command structure they found. Whereas the ANZAC at Gallipoli had been virtually an independent force due to the isolation of its operational area, I and II ANZAC on the Western Front formed but a small fraction of a British Expeditionary Force (BEF) fifty divisions strong by 1916. British commanders would control them much more closely, especially since General Sir Douglas Haig, the BEF's commander-in-chief, considered that the BEF's inexperience consequent of its own expansion no longer made "close supervision" tantamount to "interference." On the contrary, it was a "legitimate and necessary" command function.[30] Moreover, the operational responsibilities associated with command levels were changing. A 1914 corps was merely a conduit through which army orders went to divisions. By 1916, the wider spans of command resulting from the BEF's growth necessitated corps taking a major role in planning and running operations.[31]

The 5th Australian Division became the first Australian division to see serious action. Temporarily attached to the British XI Corps, it attacked at Fromelles on July 20, 1916. The operation sought to deter local German units from reinforcing the Somme, where the first great British offensive was underway. Exemplifying Haig's dictum on close supervision, XI Corps' plan was quite prescriptive, even specifying battalion frontages. It also set an objective that was overlooked by German forces.[32] Though half of his recently arrived division had just moved to the front line for the first time, McCay did not protest. He did, however, give his brigadiers flexibility in planning their own attacks, allowing them to decide when their first waves would deploy into no man's land to await the lifting of the scheduled supporting bombardment, and ordered trenches dug to the German line once it had been taken.[33] The Australians reached their objective but found their flank exposed when the British division on their right fell back. Violating the fundamentals of mission command by exceeding the authority granted him, McCay lost control of the battle, allowing battalions to be fed in contrary to corps instructions until he had virtually exhausted his division's manpower. It ultimately withdrew with heavy loss.

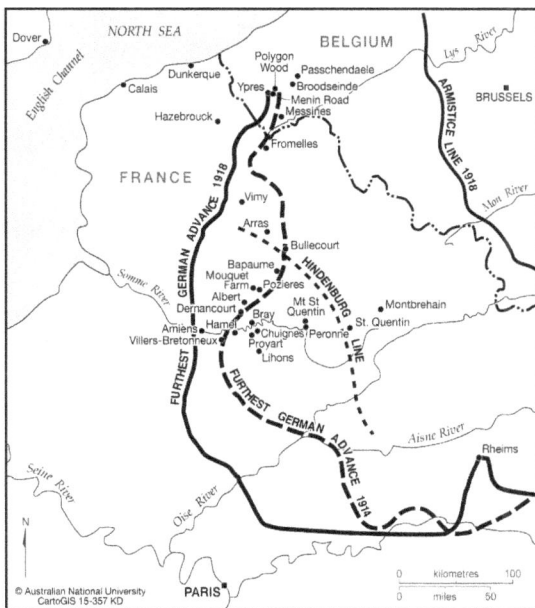

Map 2-2: Australian operations on the Western Front[34]

Somme

I ANZAC was on the Somme as part of the Reserve (later Fifth) Army commanded by British General Sir Hubert Gough. Having a well-deserved reputation for impetuosity and holding the 1914 view of corps, he preferred to bypass them and issue detailed orders directly to divisions.[35] The 1st Australian Division was not yet in the area when Gough told Walker on July 18th that it must attack Pozières the following evening, directing that the assault be made from the southwest. An experienced divisional commander, unlike McCay, Walker erupted in protest at the detailed direction and short notice. He wanted to attack from the more sheltered southeast after appropriate time for reconnaissance and other preparations. Gough relented and the attack was deferred until 23rd July.[36] Pozières quickly fell. The 2nd Australian Division relieved the 1st.

Wanting to capitalize on this success, Gough then pressured Legge, the 2nd Division's commander, to attack the nearby Pozières heights on July 29th—before the preparations were complete and despite torrential German shelling. Neither Birdwood nor White objected. Failure was total. Learning from this experience, Legge resisted Gough's impatience for a second attack until jump-off trenches could be dug to reduce the distance across no man's land. The digging had to be done under German bombardment and two grudging postponements from Gough were necessary before the heights were attacked and captured on August 4th. His approach to the Pozières battles was the antithesis of modern mission command.

Gough now ordered I ANZAC to advance on Mouquet Farm with the intention of isolating the fortress of Thiepval. The ground confined the attacks to brigade frontages. John Gellibrand, who led two Australian brigades, later remarked that a brigade commander had "little scope beyond oiling the works and using his eyes." He could do little more than decide how many of his battalions to use and how to deploy them within his assigned zone of attack.[37] But the initiative of brigade commanders—and their battalion commanders — was nonetheless prominent at Mouquet Farm. Both contributed at the planning conference for the 4th Brigade's August 8th assault. When the adjacent British assault crumpled to expose the Australian 15th Battalion's flank, the 15th's commander ordered a withdrawal. The 16th Battalion's commander opted to attack obliquely during the next assault rather than frontally as suggested.[38] He got his way. On August 26th, Gellibrand departed from the pattern of afternoon or night assaults by having the 6th Brigade attack at dawn, taking advantage of darkness to cover the assembly. Mouquet Farm held out nonetheless. Having created a salient that the Germans could shell from three sides, Gough's concept for taking it was beyond redemption.

Bullecourt

Australian brigades were at the fore when I ANZAC joined the follow-up of the German retirement to the Hindenburg Line in March 1917 after a debilitating winter on the Somme. Advancing across open country, each of the two divisions on the corps' frontage was led by a brigade-based all-arms column. The fluid nature of operations made mission command essential. Divisional commanders controlled the rate of the advance but otherwise intervened only when necessary, such as when Brigadier General Harold Elliott, the 15th Brigade's commander, sought to attack a village in defiance of a halt order.[39] Similarly, brigade commanders had to give battalions more freedom so that they could conduct the semi-independent operations needed to maintain momentum.[40]

On April 11th, the 4th Australian Division, now led by Australian Major General William Holmes, attacked the Hindenburg Line at Bullecourt as part of the Fifth Army's support for the Third Army's Arras offensive. As artillery had no time to cut the wire, Gough foisted some tanks on the 4th Division at the last minute to crush it. The Australians had never worked with tanks. Those provided were barely serviceable, but the protests of Birdwood and White were overridden. The 4th Division took the Hindenburg Line despite most of the tanks failing to reach the wire but was forced out when artillery commanders denied the infantry's calls for support, believing they had advanced further than was actually the case.[41]

Another assault at Bullecourt, on May 3rd by the 2nd Division under Major General Nevill Smyth, a British officer who had replaced Legge, was successful. As only one division was involved, the remainder of I ANZAC had little to do; Smyth and his staff handled the preparations. The respective divisional and corps roles were repeated when the 1st and 5th Divisions were fed in. Gellibrand showed how a brigade commander could influence the battle if he were far enough forward. Siting his headquarters almost on the front line, he intervened at crucial stages, such as when the 28th Battalion mistakenly began to withdraw. Nonetheless, Gellibrand was unable to dissuade Smythe, located far to the rear, from ordering an assault that he realized would be suicidal.[42] The failure of higher commanders to grant their subordinates authority in keeping with their better grasp of battle conditions threatened success and unnecessarily cost Australian lives in both Bullecourt battles.

Messines

Meanwhile, the 3rd Australian Division—under Monash and belonging to II ANZAC—was preparing for its first big fight since its arrival in France at the end of 1916: Second Army's June 7, 1917 attack on Messines Ridge. Second Army's commander, General Sir Herbert Plumer, tended to be hands-off, shaping plans in consultation with his corps commanders and encouraging them to do likewise with their subordinates.[43] Aware that he had two outstanding divisional commanders in Monash and the New Zealander Russell, II ANZAC's commander, Godley, followed

Plumer's advice. Monash's task was a complex one for an untried formation, involving as it did a complicated turning maneuver. Monash therefore favored prescription over more general guidance in dealing with his subordinates. He told his brigade commanders how they were to employ their battalions and, in some cases, their platoons. Still, his guidance was not one-way. Brigade commanders, heads of staff branches, and frequently battalion commanders attended conferences at which Monash sought their opinions and amended his orders as he thought appropriate based on their advice.[44] The 3rd Division performed superbly in the attack, which seized Messines Ridge and thereby secured communications into the Ypres Salient.

Third Ypres

The Germans overlooked the Ypres Salient and the BEF launched the Third Ypres offensive at the end of July to evict them. Haig again turned to Plumer's army and its two ANZAC when Gough's Fifth Army stalled on a battlefield turned to mud by torrential rain. Plumer's plan was based on the "bite and hold" approach: a limited advance behind a heavy "creeping" barrage, the advance halting before resistance hardened. Corps' plans adopted this guidance and, for the first assault, on the Menin Road on September 20, 1917, conformed to I ANZAC's scheme as it had the main role in Second Army.[45] Divisions assaulted side-by-side, each on a narrow frontage so that fresh troops were available in strength for each phase of the assault. Benefiting from much improved weather, the attack succeeded, as did the subsequent ones on Polygon Wood and Broodseinde, which employed the same tactics. Taking advantage of the independence granted division commanders, Monash added additional phases to those in the II ANZAC plan for his division's assault at Broodseinde.[46] Australian Major General Talbot Hobbs had taken over the 5th Division from

McCay. He gave the 15th Brigade under Elliott a free hand to restore the situation after a German spoiling attack threatened the Polygon Wood assault at the last minute.[47] Only the return of rains derailed a final attack involving I and II ANZAC at Passchendaele on October 12, 1917.

Figure 2-2: Australian soldiers near the Menin Road in what was once Chateau Wood during the Third Ypres Campaign in 1917[48]

The German Spring Offensives

I and II ANZAC ceased to exist with creation of the Australian Corps in November 1917. Grouping the five Australian divisions together enhanced the trust and teamwork we consider fundamental to today's conceptualization of mission command. Unfortunately, the newly formed corps could not take full advantage. During the emergency created by the German offensives in spring 1918, it was required to fight piecemeal by division, and even brigade, within British formations. British VII Corps commander, Lieutenant General Sir Walter Congreve, employed the 3rd and 4th Australian Divisions to fill gaps that exposed the communications center of Amiens during fighting on the Somme. Owing to the urgency, his orders on March 27th to Monash and Sinclair-MacLagan (who had succeeded Holmes as Commander, 4th Division when Holmes was killed after Messines) were necessarily brief. Later working over a map by candlelight, Monash scribbled instructions in which "each phase of action was clearly explained and the action then crisply ordered."[49] Dispatch riders took the orders to the brigades moving to join him; their commanders oversaw the deployment of the battalions between the Somme and the Ancre next morning. Monash had left the detailed execution to them. Sinclair-MacLagan likewise did not fetter his brigadiers as they deployed their brigades at Dernancourt. There, too, fluid battlefield conditions precluded close control. Platoons and sections, now led by vastly experienced commanders, used their initiative in these conditions to seize German posts opposite them by using stealth, a process that was dubbed "peaceful penetration."

Contrasting approaches to mission command were evident during separate operations at Villers-Bretonneux, the town shielding Amiens. Villers-Bretonneux was defended by the 9th Australian Brigade (attached to the British 18th Division at the time) when the Germans reached its outskirts on April 4th. So swift was the German assault that detailed guidance was impossible. Lieutenant Colonel Goddard of the 35th Battalion, which held the town, only had time to tell Lieutenant Colonel Milne of the 36th, which had come forward from reserve, to "counterattack at once." Milne could only give his company commanders their frontages before ordering, "Go till you're stopped and hold on at all costs."[50] The town was held.

On April 24th the town was lost when the Germans, using tanks for the first time, wrested it from the British 8th Division under Major General William Heneker. Though Elliott's 15th Brigade was well positioned to counterattack immediately, Heneker tasked his own division, unaware that it had been shattered. Its counterattack failed. General Sir Henry Rawlinson, who was responsible for Villers-Bretonneux as the commander of the Fourth Army, dispatched the 13th Australian Brigade under Brigadier General William Glasgow. On receiving detailed instructions on how to attack from Heneker, Glasgow retorted, "Tell us what you want us to do so, sir, but you must let us do it our own way."[51] Heneker yielded. Glasgow's 13th and Elliott's 15th Brigades simultaneously carried out an enveloping movement, many subordinate commanders having to give orders while on the move. Villers-Bretonneux was regained.

Monash and Hamel

Lieutenant General John Monash replaced Birdwood as Australian Corps commander at the end of May 1918. The holder of multiple academic degrees and a pioneer of reinforced concrete construction in Australia, Monash dwarfed Birdwood in terms of intellect and vision. White, Birdwood's chief of staff, could not recall Birdwood ever drafting a plan; the same could not be said of Monash.[52] According to Gellibrand, Monash regarded his operations chief as a clerk and urged him, as the 3rd Division's new commander, to draft his own orders.[53] Meticulously choreographed, Monash's plans were "like nothing so much as a score for an orchestral composition" in which each arm and unit contributed to the general harmony by playing its part.[54] The divisional commanders were now proven: Glasgow (1st Division), Major General Charles Rosenthal (2nd), Gellibrand (3rd), Sinclair-MacLagan (4th), and Hobbs (5th). They reflected the expertise of the BEF in 1918, a situation that allowed higher commanders to apply a light touch to the tiller.[55]

With the German tide ebbing, Monash enjoyed much more freedom of action than Birdwood had.

Figure 2-3: Australian and American soldiers share a trench above the newly captured village of Hamel, July 4, 1918[56]

Rawlinson, Monash's army commander, wanted the village of Hamel taken to gain more room for the defense of Villers-Bretonneux. Monash agreed provided he received some of the latest tanks. Rawlinson gave him sixty and left the planning entirely to him. As the tanks would lead the attack with the infantry following, Monash dispensed with the creeping barrage so that his armor could forge ahead unimpeded. But he had not reckoned on the legacy of the disastrous Australian experience with tanks at Bullecourt. Sinclair-MacLagan and the brigadiers in Monash's composite infantry force (put together to avoid the losses falling entirely on one division) all urged the retention of the creeping barrage. As they would be the men on the spot, Monash deferred even though it meant abandoning his original concept.[57]

Tanks and infantry, which included some American troops, would advance together. Monash also subordinated tank commanders to their infantry counterparts against the armor leaders' protests. All concerned thrashed out the details of the July 4 attack, particularly the intricate coordination of aircraft, infantry, tanks, and artillery, at planning conferences. Hamel fell in ninety-three minutes.

Advancing to Victory

Hamel was a blueprint for the bigger set-piece battles to come. They started with the Fourth Army's advance on August 8th during which the Australian and Canadian Corps, attacking side-by-side south of the Somme, had the principal roles. Rawlinson imposed "no limitations or conditions" apart from setting frontages and phase lines.[58] Monash requested he expand the first phase to include overrunning the German gun line in order to prevent the enemy's withdrawing their artillery. His arrangements for the Australian Corps blended prescription and devolution. Seeking to shorten the approach marches of the divisions destined for the final objective, he positioned them closest to the start line, where the two divisions taking the first objective would leapfrog them. They would then leapfrog those divisions on the first objective and go on to the final one.

The extraordinarily complex scheme had never been attempted before. Monash trusted to the competence and intelligence of his commanders to carry it out.[59] His division commanders coordinated the preliminary movements behind the line of departure amongst themselves, Monash telling them he "would have nothing to do with it" beyond his initial guidance.[60] While the first phase of the attack was a Hamel-like set-piece, he envisioned the second phase, which would go beyond field gun range, as an open warfare advance that devolved mainly onto brigades.[61] As well as tanks, Monash allotted extra resources, especially artillery and engineers, to the brigades involved, creating all-arms brigade groups that gave their commanders the means to undertake a deep penetration without a creeping barrage. The Australians and Canadians, supported by the British, swept all before them on August 8, 1918, which the Germans called their army's "black day."

Figure 2-4: Australian troops and supporting tanks wait for the start of the second phase of the great attack, August 8, 1918[62]

A new challenge immediately arose. Used to years of static warfare, commanders now had to make the transition to mobile conditions and more decentralized control. The start of the Australian advance on August 9, 1918 was badly rushed due to Rawlinson's failure to ensure that the Fourth Army's coordinating instructions were promulgated in good time. Monash stressed the need for "close liaison" between his divisional commanders to ensure their formations moved together, but they did not carry it out. The 2nd Division began its assault much later than the 1st Division. Next day Monash was "hands on," personally briefing the brigadiers whose brigades would execute his plan to pinch out the U-shaped bends in the Somme at Etinehem and Proyart. As the plan was breathtakingly bold, he probably did not want it conveyed through the divisional commanders, though they were present at the briefing.[63] He twice meddled with the 1st Division attack on Chuignes on August 23rd, but allowed Gellibrand to organize the 3rd Division's envelopment of Bray on the 24th. Monash's command of the Australian Corps was starting to reflect the trend in the BEF at this stage of the war of passing as much control down to divisions as possible.[64] To this end, he allocated each division a corridor along which it moved on a frontage determined by its commander—invariably a brigade frontage. Lead battalions were given two field guns for quick destruction of pockets of resistance.

The Canadian Corps had now departed Fourth Army, leaving the Australian Corps as its sole spearhead. Rawlinson gave Monash free rein, even turning a blind eye when Monash, determined to give the Germans no respite, circumvented orders to halt and attacked Mont St Quentin and the town of Péronne on the great Somme bend.[65] After a frontal assault on August 28th failed because the Somme crossings were untenable, he orchestrated a wide flanking wheel that utilized crossings behind the Australian line. One of the Western Front's rare maneuver battles, involving the 2nd, 3rd, and 5th Australian Divisions, resulted. It moved so quickly that orders were soon outdated, making mission command based on Monash's intent critical. Tactics were "largely left to divisional, brigade, battalion and even platoon commanders."[66] When the 2nd Division's lead brigade was unable to cross the Somme at Halle on August 29th, Rosenthal ordered it to head for Ommiécourt further west. Finding that crossing likewise unusable, he directed it on to Feuillères even further to the west. Once informed of Rosenthal's decision, Monash postponed the assault for twenty-four hours.

In the ongoing fight for Péronne, too, Monash's intent was often the only guidance available. Failing in his attempt to hustle the 15th

Brigade over the Somme near the town, Elliott was unable to co-ordinate with the 14th Brigade the renewed attack planned for September 2nd. Sketchy orders ensued, requiring battalion commanders to act on their own judgement. Consequently, the 58th Battalion ended up clearing more of Péronne than it was supposed to in the original plan.

Figure 2-5: Lieutenant General Sir John Monash in 1918[67]

Monash again gave his division commanders corridors along which to advance in the pursuit that took Fourth Army to the Hindenburg Outpost Line. Rawlinson, Monash, and the commanders of III and IX Corps planned the attack on it "quite informally over a cup of afternoon tea" on September 13th.[68] Of the planning within the Australian Corps, Monash wrote that by now commanders understood each other and the tactical methods to be used so well that they could anticipate the action required of them. Much of what he had to say "could almost be taken for granted."[69] It was a maturation of trust and competence born of years of familiarity and the harsh experience of war. Attacking the outpost line on September 18, 1918, 1st Division easily took its objective. The 4th Division was held up by stiffer resistance. Sinclair-MacLagan, who had earlier left the tactics largely to his brigades, ordered a renewed assault on convergent axes. It succeeded. The action was to prove the two divisions' last battle of the war. Rawlinson assigned Monash the untried American 27th and 30th Infantry Divisions to replace them for the September 29th assault on the Hindenburg Line.

As the Australian Corps would have the main role, Rawlinson called on Monash to shape the Fourth Army's plan. Monash feared a costly assault across the St. Quentin Canal, which formed part of the Hindenburg Line. He argued the attack should instead pass over the tunnel through which the canal ran in the Australian sector, with the British III and IX Corps following the Australians and Americans to extend the breach after the crossing.[70] Rawlinson disagreed, and included a canal assault.[71] His decision paid off. The Americans' set-piece attack over the tunnel stalled. Knowing that their task was daunting and their experience limited, Monash had taken a prescriptive approach to them, mentoring their commanders and attaching Australians as advisers at every command level down to battalion. It was not enough. The Americans were pinned down, holding up the trailing 3rd Division that was to undertake the more complex open warfare advance with the 5th Division in the operation's second phase. Commanding from his headquarters well to the rear, Monash rejected the advice of the commanders on the spot and ordered a renewed advance. It failed. But the British 46th Division successfully crossed the canal on the right of the Americans, justifying Rawlinson's decision to assault on a wide frontage. That success helped the American 30th Division to reach the tunnel line, creating a breach through which Monash swung the Australian assault next day. The attack on the Hindenburg Line was the Australian's last major battle of the war.

SINAI/PALESTINE

While Australian infantry fought on the Western Front, Australian light horse fought the Turks in the Sinai/Palestine theater where, alongside New Zealand mounted troops, they formed the spearhead of the British-led Egyptian Expeditionary Force (EEF). Combat never approached the intensity of that in France and Flanders. Gains of tens of kilometers at little cost were common, the horsemen providing the mobility that the distances demanded. The combination of mobility, distance, and poor communications made decentralized command the rule rather than the exception in the Anzac Mounted Division under Harry Chauvel and, later, the Desert Mounted Corps, which he went on to lead. This was especially the case at the regimental level where "quick decisions had to be made on the move, sometimes at crucial moments. A gallop in the wrong direction could soon lead to disaster."[72] It was Lieutenant Colonel George Bourne, the commander of the 2nd Light Horse (LH) Regiment, who chose when his unit should pull back from its outpost line during the July 1916 battle of Romani, in which a Turkish advance across the Sinai was checked. The EEF now took the offensive. At the end of 1916, its light horse spearhead had cleared the Sinai and reached southern Palestine.

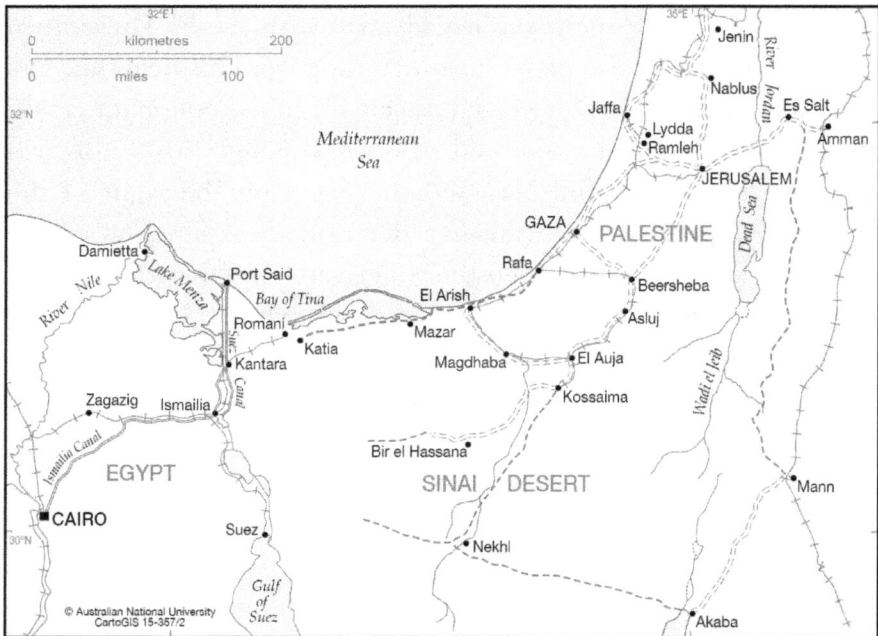

Map 2-3: The Egypt and Palestine theater, 1916-1918[73]

By then Chauvel's subordinates were highly experienced and he knew them well. At times they disobeyed his direct orders in light of their understanding of both his overarching intentions and their more intimate knowledge of the tactical situation. Worried by a water shortage at Magdhaba, Chauvel ordered a withdrawal just as the 1st LH Brigade was about to attack in December 1916. Its commander, Brigadier General Frederick Cox, ignored the order, the enemy folding during the ensuing assault. Seeing opportunity in the absence of specific orders, the 10th LH Regiment's commander, Lieutenant Colonel Horrie Robertson, attacked the Turkish rear and blocked their retreat.[74] Water shortages and knowledge of approaching Turkish reinforcements prompted another withdrawal order from Chauvel at Rafa, which three brigades had struck from different directions in January 1917. Again Chauvel's subordinates ignored the order, pressing their assaults and forcing the garrison's surrender.[75]

The Turkish defense of southern Palestine hinged on Gaza. The EEF's set-piece plan to take it in March 1917 dictated a frontal attack by British infantry while Chauvel's mounted force enveloped the town. Ordered to capture Gaza when the infantry attack failed, Chauvel hurriedly redeployed his brigades to assault from two directions, leaving the tactical details to their commanders. The troopers were already in Gaza when the British high command ordered a withdrawal in the face of an approaching Turkish formation.

Under the vigorous leadership of General Sir Edmund Allenby, the EEF attacked Gaza again in October. While a feint against the town preoccupied the Turks, the main assault, which included the Desert Mounted Corps, struck more lightly held Beersheba at the far end of the enemy line. With daylight fading and the Turks still in possession of Beersheba's all-important wells, Chauvel gave a five-word order that made both mission and intent clear: "Put Grant straight at it."[76] Brigadier General William Grant's 4th LH Brigade launched a spectacular charge that took the wells. Gaza fell a few days later. Allenby entered Jerusalem in December.

Figure 2-6: Australian Light Horse on their way to attack Beersheba at the end of October 1917[77]

Though April and May 1918 operations against Turkish lines of communications on the Jordan flank were unsuccessful, they distracted the Turks from the coastal flank where Allenby intended to break through on ground more suited to offensive operations. The Australian and New Zealand horsemen led the pursuit that followed his great victory at Megiddo in September. Here too commanders exercised initiative and adapted orders to capitalize on opportunity. Chauvel chose to advance through the vast breach torn in the Turkish line on two axes instead of one as Allenby preferred.[78] The advance itself—left in the hands of brigade and regimental commanders—featured flanking maneuvers and envelopments. It culminated for the Australians at Damascus in October.

LOOKING BACK

Like the British Army, the pre-World War I Australian Army was aware of what would now be considered the fundamentals of mission command. But there were few opportunities to develop anything akin to a mission command culture. Untried when war came, AIF commanders had to learn on the job, and their early attempts at decentralized operations had mixed results. The Anzac landing failed in large part because subordinates lost sight of their superiors' intent. In the August offensive at Anzac, Birdwood made sure that his intent was clear at Lone Pine, leaving detailed planning to Walker. Yet elsewhere Monash had to break free from the shackles imposed by Cox during the march on Hill 971. On the Western Front, I and II ANZAC and, later, the Australian Corps had to deal with contrasting British command approaches. Gough's command style in 1916 was "hands on;" Plumer in 1917 and Rawlinson in 1918 gave their subordinates plentiful latitude. Mirroring the trend in the wider BEF, the input of Australian division and brigade commanders into their superiors' plans generally increased as the war went on. Monash routinely met with his commanders to solicit their input before a battle, providing opportunities for him to get their ideas and for them to ensure that they understood his intent. But Bullecourt and Villers-Bretonneux showed that deference to the commanders on the spot was never automatic. In the Sinai and Palestine, however, terrain, distance, limitations of communications technology, and the inherently decentralized character of mounted operations meant mission command was virtually essential.

Experience, trust, and personality underpinned mission command in all three theaters. Bridges gave Sinclair-MacLagan the mission of leading the assault at the Anzac landing because he was the most experienced of his brigadiers, which engendered more trust in his abilities. Monash took a prescriptive approach at Messines because it was his division's first major battle. In 1918 the Australian Corps was a well-proven formation and Monash's ability was well established, enabling Rawlinson to give him a significant role in Fourth Army's operational planning. It has to be said in fairness to Gough that he could not regard I ANZAC's commanders in the same light two years previously.

As corps commander, Monash often immersed himself in detail more than was justified. This might be attributed to personality; he had an engineer's mind for specifics, method and, more importantly, he towered over his colleagues cerebrally. Nonetheless, Rawlinson did not always defer to him, intervening when he thought it necessary as was the case during Monash's planning for the Hindenburg Line. Post-war editions of *Field Service Regulations* recognized the importance of balance in the application of what we today call mission command. Unlike prewar editions, these recognized that mission command did not dictate a "hands off" approach regardless of subordinates' experience, familiarity, and the situation at hand, but rather application of its elements in a "happy medium."[79]

ENDNOTES

1   General Staff, *Field Service Regulations, Part 1, Operations*.
    London: War Office, 1909, 27.

2   Chris Pugsley, "We have been here before: the evolution of
    the doctrine of decentralised command in the British Army
    1906–1989". Sandhurst Occasional Papers No. 9, Royal Military
    Academy Sandhurst, 2011, 11.

3   General Staff, *Infantry Training (4 Company Organization) 1914*,
    London: HMSO, 1914, 121.

4   John Monash, "Lessons of the Wilderness Campaign 1864" in
    Commonwealth Military Journal, April 1912, 276.

5   John Monash, *100 hints for company commanders*, 3DRL/2316,
    Series 3/3, Australian War Memorial (AWM).

6   Albert Palazzo, *The Australian Army*. South Melbourne:
    Oxford, 2001, 47; Alec Hill, *Chauvel of the light horse*. Carlton:
    Melbourne University Press, 1978, 54.

7   Peter Pedersen, *Monash as military commander*. Carlton:
    Melbourne University Press, 1978, 35.

8   Charles Bean, *Official History of Australia in the war of 1914–18.
    I. The story of Anzac*. Sydney: Angus and Robertson, 1921, 54.

9   Eric Sixsmith, *British generalship in the twentieth century*.
    London: Arms and armour, 1970, 153.

10 General Headquarters MEF War Diary, Force Order No. 1 and
    sketch, and Instructions for GOC A&NZ Army Corps, all dated
    April 13, 1915, AWM4 1/4/1/2, Part 2.

11   Map courtesy of the Australian National University CartoGIS, Adapted from "Gallipoli 12-123 Map 5."

12   Photo from the Australian War Memorial photograph collection, item number P10140.008, https://www.awm.gov.au/search/all/?query=p10140.008&op=Search&format=list&rows=20&filter%5Btype%5D=Photograph&section%5B0%5D=collections (accessed December 18, 2015).

13   Birdwood, "Instructions to GOC 1st Australian Division, 18 April 1915" in Charles Aspinall-Oglander, *Official History of the Great War. Military Operations. Gallipoli, Maps and Appendices, I.* London: Heinemann, 1929, 42; Robin Prior, *Gallipoli. The end of the myth.* Sydney: University of New South Wales Press, 2009, 110–11.

14   Chris Roberts, *The landing at Anzac 1915.* Sydney: Big Sky, 2015, 106–7.

15   Ibid., 100-101.

16   Bean, *Story of Anzac, I,* 364–5. Bean asserts (p. 359) that Sinclair-MacLagan intended to resume the advance when the 2nd Brigade arrived.

17   Peter Pedersen, *Anzac treasures. The Gallipoli collections of the Australian War Memorial.* Sydney: Murdoch Books, 2014, 109.

18   Chris Roberts, "The Landing at Anzac: a re-assessment" in *Journal of the Australian War Memorial,* April 1993, 31.

19   Roberts, *Landing at Anzac,* 100–1 and 134.

20   Bean, *Story of Anzac, I,* 455.

21 Edward Erickson, *Gallipoli. Command under fire*. Oxford: Osprey, 2015, 132.

22 Godley to Monash, 27 April 1915, 3DRL/2316, Series 3/10, AWM. (Emphasis in original)

23 Pedersen, *Monash as military commander*, 86.

24 Charles Bean, *Official History of Australia in the war of 1914–18. II. The story of Anzac*, Sydney: Angus and Robertson, 1924, 452–3.

25 Peter Burness, *The Nek. A Gallipoli tragedy*. Auckland, Exisle Publishing, 2012, 108, 113.

26 Charles Aspinall-Oglander, *Official History of the Great War. Military Operations. Gallipoli, II*. London: Heinemann, 1932, 192.

27 Pedersen, *Monash as military commander*, 108, 112–13.

28 Charles Bean, *Official History of Australia in the war of 1914–18. III. The AIF in France: 1916*. Sydney: Angus and Robertson, 1929, 45–6.

29 Charles Bean, *Anzac to Amiens*. Canberra: Australian War Memorial, 1968, 190.

30 Peter Simkins, "Haig and his army commanders". In *Haig: a reappraisal 70 years on*, edited by ed. Brian Bond and Nigel Cave, Barnsley: Leo Cooper, 1999, 94.

31 Andy Simpson, *Directing operations. British corps command on the Western Front 1914–1918*. Port Stroud: Spellmount, 2006, 19, 29–30.

32 XI Corps order 57, 15 July 1916, AWM4 1/22 Part 2.

33 5th Australian Division War Diary, 'Instructions for brigadiers' 15 July 1916 and Order 5, 16 July 1916, AWM4 1/50/5.

34 Map courtesy of the Australian National University CartoGIS, "Western Front.pdf."

35 Andy Simpson, "British corps command on the Western Front 1914–1918". In *Command and Control on the Western Front*, edited by Gary Sheffield and Dan Todman, Staplehurst: Spellmount, 2004, 204 and 106; Andrew Farrar-Hockley, *Goughie*. London: Hart Davis MacGibbon, 1975, 188.

36 Gary Sheffield, *Command and Morale. The British Army on the Western Front 1914–1918*. Barnsley: Praetorian Press, 2014, 63.

37 Gellibrand in Peter Sadler, *The paladin. A life of Major General Sir John Gellibrand*. South Melbourne: Oxford, 2000, 85.

38 Bean, *The AIF in France 1916*, 740.

39 Ross McMullin, *Pompey Elliott*. Carlton North: Scribe Publications 2010, 273–3.

40 Robert Stevenson, *To win the battle. The 1st Australian Division in the Great War, 1914–1918*. Port Melbourne: Cambridge University Press, 2010, 165.

41 Charles Bean, *Official History of Australia in the war of 1914–18. IV. The AIF in France: 1917*. Sydney: Angus and Robertson, 1933, 336–7.

42 Sadler, *The paladin*, 132.

43 Geoffrey Powell, *Plumer. The soldier's general*. London: Leo Cooper, 1918, 171.

44 Pedersen, *Monash as military commander*, 168; see also Geoffrey Drake-Brockman, *The turning wheel*. Perth: Patterson Brokensha, 1960, 117.

45 Bean, *The AIF in France 1917*, 735.

46 Peter Pedersen, *The Anzacs. Gallipoli to the Western Front*. Camberwell: Penguin Viking, 2007, 257; see also Simpson, *Directing operations*, 109.

47 Bean, *The AIF in France 1917*, 809–10.

48 Photo from the Australian War Memorial photograph collection, item number E01220, https://www.awm.gov.au/search/all/?query=eo1220&submit=&op=Search&format=list&section%5B%5D=events&section%5B%5D=units&section%5B%5D=places&section%5B%5D=articles&section%5B%5D=books&section%5B%5D=people&section%5B%5D=collections (accessed December 18, 2015).

49 Charles Bean, *Official History of Australia in the war of 1914–18. V. The AIF in France: during the main German offensive 1918*. Sydney: Angus and Robertson, 1937, 177.

50 RSM A. Horwood in Bean, *The AIF in France: during the main German offensive*, 340.

51 Glasgow in Bean, *The AIF in France: during the main German offensive*, 574–5. For a full version of this conversation see Peter Edgar, *Sir William Glasgow*. Newport: Big Sky, 2011, 203–5.

52 White in Bean diary 113, 30 May 1918, AWM38 3DRL 606/113/1.

53 Gellibrand's comments on Monash's *Australian victories*, 3DRL/1473, [91], AWM.

54 John Monash, *The Australian Victories in France in 1918*, London: Hutchinson, 1920, 56.

55 Robin Prior and Trevor Wilson, *Command on the Western Front*. Oxford: Blackwell Publishers, 1992, 305.

56 Photo from the Australian War Memorial photograph collection, item number E02844A, https://www.awm.gov.au/search/all/?query=e02844a&submit=&op=Search&format=list&section%5B%5D=events&section%5B%5D=units&section%5B%5D=places&section%5B%5D=articles&section%5B%5D=books&section%5B%5D=people&section%5B%5D=collections (accessed December 18, 2015).

57 Peter Pedersen, *Hamel*. Barnsley: Pen and Sword, 2003, 48–9.

58 Monash to Bruche, October 10, 1918, MS1884, Series 1B, National Library of Australia

59 Monash, *Australian victories*, 95.

60 Charles Bean, *Official History of Australia in the war of 1914–18. VI. The AIF in France: May 1918–the armistice*. Sydney: Angus and Robertson, 1942, 493.

61 Monash, 'Allotment of infantry to objectives', c. 23 July 1918, 3DRL/2316, Series 3/62, AWM.

62 Photo from the Australian War Memorial photograph collection, item number E03883, https://www.awm.gov.au/search/all/?query=e03883&submit=&op=Search&format=list&section%5B%5D=events&section%5B%5D=units&section%5B%5D=places&section%5B%5D=articles&section%5B%5D=books&section%5B%5D=people&section%5B%5D=collections (accessed December 18, 2015).

63   Bean, *The AIF in France: May 1918–the armistice*, 687.

64   Simpson, *Directing operations*, 162.

65   Prior and Wilson, *Command on the Western Front*, 341–2.

66   Bean, *The AIF in France: May 1918–the armistice*, 873.

67   Photo from the Australian War Memorial photograph collection, item number A02697, https://www.awm.gov.au/search/all/?query=a02697&submit=&op=Search&format=list&section%5B%5D=events&section%5B%5D=units&section%5B%5D=places&section%5B%5D=articles&section%5B%5D=books&section%5B%5D=people&section%5B%5D=collections (accessed December 18, 2015).

68   Monash, *Australian victories*, 221.

69   Ibid.

70   The plan is reproduced in full in *Australian victories*, 236-40; for Monash's arguments, see 'Plan for Beaurevoir offensive', 21 September 1918, 3DRL/2316, Series 3/65, AWM.

71   John Harris, *Amiens to the armistice. The BEF in the Hundred Days' Campaign, 8 August–11 November 1918*. London: Brassey's, 1998, 206, 210–12.

72   Jean Bou, *Light horse. A history of Australia's mounted arm*. Port Melbourne: Cambridge University Press, 2010, 212.

73   Map courtesy of the Australian National University CartoGIS, "035 Palestine."

74   Henry Gullett, *Official History of Australia in the war of 1914–1918. VII. Sinai and Palestine*. Sydney: Angus and Robertson,

1944, 221–3.

75 Hill, *Chauvel*, 93.

76 Gullett, *Sinai and Palestine*, 393.

77 Photo from the Australian War Memorial photograph collection, item number A02788, https://www.awm.gov.au/search/all/?query=a02788&submit=&op=Search&format=list&section%5B%5D=events&section%5B%5D=units&section%5B%5D=places&section%5B%5D=articles&section%5B%5D=books&section%5B%5D=people&section%5B%5D=collections (accessed December 18, 2015)

78 Alec Hill, "General Sir Harry Chauvel". In *The commanders*, edited by David Horner, North Sydney: Allen and Unwin, 1992, 79.

79 General Staff, *Field Service Regulations, Part 1, Operations*. London: War Office, 1924, 10.

3

MISSION COMMAND IN WORLD WAR II: AUSTRALIA, MACARTHUR'S GENERAL HQ AND THE SOUTHWEST PACIFIC AREA

Dr. Peter J. Dean

On Wednesday, August 4, 1943 the front page of the *Brisbane Courier Mail* featured the progress of the battle for Sicily in the Mediterranean. While a vital and critical part of the war, not many Australians troops were in action in the Mediterranean at the time. The more geographically relevant story for the local audience was a much smaller article tucked midway down the left hand side of the front page. It described the establishment of artillery positions within range of the town of Salamaua in New Guinea. The article went on to cover the constant air bombardment of Salamaua and positioning of Australian and US ground troops outside of the town.[1] Unknown to the Australian public, what the story was actually detailing was the final stage of the major Allied deception operation in 1943 for New Guinea. The 3rd Australian Division, supported by a regiment from the US 41st Infantry Division, had been closing on Salamaua for months, drawing more and more Japanese troops away from the real objective: the town of Lae located north of Salamaua at the mouth of Huon Gulf.

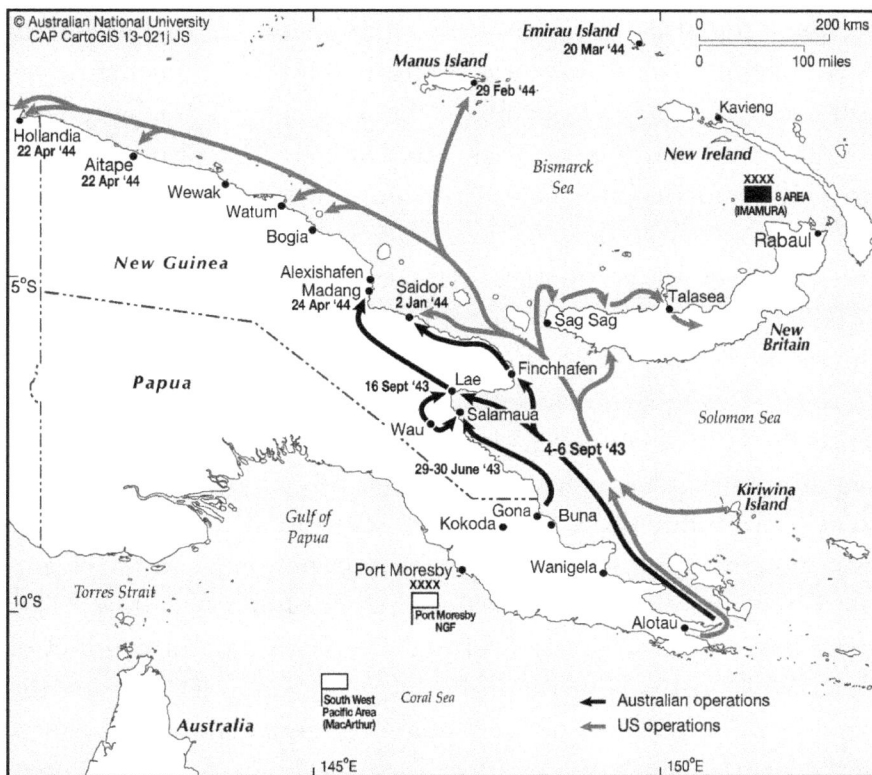

Map 3-1: Context for 1943 Australian-US operations against Lae and Finschhafen[2]

Lae represented the hinge of the Japanese defensive positions in New Guinea. In just over three weeks, Australian and US forces in General Douglas MacArthur's Southwest Pacific Area (SWPA) would launch an all-out attack on Lae. The US Army's 503rd Parachute Infantry Regiment would land on the abandoned Nadzab afield of just outside of Lae, securing it for follow-on air landing of the 7th Australian Division. The 9th Australian Division would make an amphibious landing to the west of Lae the day before the 7th Division's assault from the air. Once established in their respective bridgehead and airhead, the two divisions would advance on and destroy the Japanese base.[3]

Planning and organization for this operation, the largest in Australian military history, was far from easy. While ground forces were overwhelmingly Australian (the 3rd, 7th, and 9th Divisions in the diversion and the initial assault, with the 5th and 11th divisions as reinforcements), the majority of supporting assets – most of the naval power, amphibious landing craft, and a significant portion of the air power – was American. The American role was even more significant in terms of command relationships: the operation was under the command of US Army General Douglas MacArthur's General Headquarters Southwest Pacific Area (GHQ SWPA). Though the presence of a single theatre-level commander made organization and command lines relatively straight forward, the joint and multinational nature of the coalition meant there was also potential for considerable tension between Australia and the US, especially in terms of doctrine and military culture. This potential was soon realized during the planning and conduct of the assault on Lae (Operation Postern). The contrast between the Americans' highly centralized planning and command systems and decentralized Australian (British) system soon became sticking points for leaders from both countries.[4] The Australians feared what they perceived to be a US obsession with running everything centrally from GHQ SWPA. It was not long before the Australians felt their preference for decentralized command – a system akin to today's mission command notion of "assigning a subordinate commander a mission without specifying how the mission is to be achieved"– would be stifled by MacArthur and his staff.[5]

Such differences were never more evident than on August 4, 1943, when US Army Major General Stephen Chamberlin, the G-3 senior operations officer at GHQ, met with Major-General Frank Berryman, Deputy Chief of the General Staff (DCGS) and senior operations officer of the Australian Army in the SWPA. The American Chamberlin probed Berryman on the details of the plans for 7th and 9th Australian Divisions' assault on Lae. Berryman passed on to Chamberlin that he did not have much to add to the brief note that his commander, General Thomas Blamey, had sent to MacArthur a few days earlier. Berryman went on to say that he did not expect detailed plans to be finalized until "about ten days before D-Day when Gen Blamey would be in New Guinea." The Australian explained to Chamberlin that a directive had been given to I Australian Corps regarding the conduct of the operation and that the two assault divisions had been ordered to cooperate with their respective air and amphibious assault partners to work out the details "in accordance with the general outline plan as submitted." Both the directive and devolution of planning to lower headquarters were as per the Australian Army's doctrine.[6] Chamberlin replied that if the Australian HQ (Landops) plan had been submitted to the US Army Command and General Staff School it would have gotten "no more than [a mark of] 20 per cent."[7]

Figure 3-1: General Thomas Blamey (left) with General Douglas MacArthur[8]

Chamberlin left the meeting far from impressed. He wrote to his boss the following day, explaining to the GHQ Chief of Staff Lieutenant General Richard Sutherland, that the "submission of additional detail to GHQ" from Blamey's HQ "relating to the New Guinea Force plans was not contemplated." Chamberlin also noted that he was left with the impression that Berryman knew nothing of the detailed planning for the pending operation at Lae other than that "he was confident it was progressing well."[9] Clearly unhappy, Chamberlin recommended to Sutherland that what was required was "'on-the-spot' coordination with GHQ representatives via a visit directly to General Herring's [I Australian Corps] headquarters in Port Moresby."[10]

Hearing of the proposal, Berryman ensured such a meeting did not take place. To the Australians, such a fragrant abuse of the chain of command would only result in further meddling by MacArthur's headquarters. Berryman had played dumb with Chamberlin. In actuality, he knew full well the details of Australian plans for the assault, but Berryman and Blamey had agreed that they had had enough of GHQ interference during the earlier 1942 Papuan campaign and the conduct of deception operations in front of Salamaua over the previous few months. They were determined that MacArthur and his staff would not interfere in their plans for the Lae assault.

The reason for the differences in approach and priorities between GHQ and HQ Landops were not lost on Berryman. The problem could be explained in simple terms: "the difference is we [Australians] work on a decentralised [planning] basis where GHQ have a highly centralised one."[11] This fact was also not lost on Chamberlin, who in the previous campaign had noted, "we constantly have to deal with our allies. Their systems, their methods, and their line of thought are different from ours.... Great patience is necessary. It's behoved all of us to know when to give in and when to be firm."[12]

COALITIONS AND ALLIANCES: AN AUSTRALIAN WAY OF WAR

Since the 1788 establishment of a convict colony at Sydney Cove, Australia has always relied on an alliance with a "great and power friend" for its security. The maintenance of these asymmetrical alliances, first with Great Britain and then with the United States, has been a dominating feature of Australian strategic culture. This alliance-based culture means that Australia's military forces have almost always fought in coalitions and, with the exception of the East Timor intervention in 1999 and post-Cold War peacekeeping operations in the South Pacific, Australia has assumed the mantle of a subordinate power. Development and execution of mission command in Australia has been influenced by the country's role within these asymmetrical alliances and coalitions. This history regarding Australian use of armed force also means that its military has more often than not functioned at the strategic and tactical levels while rarely commanding at the operational level of war.

As such, the focus of this chapter is Australia's mission command approach during operations in a coalition environment. The Pacific rather than European theater provides the examples used. It was in the SWPA that the majority of Australia's military power was concentrated during World War II, and it is the only time in Australian history that the continent was directly attacked and threatened with invasion by a foreign military power. It is important to note that, unlike the deep abiding alliance partnership that exists between the US and Australia today, in the SWPA during 1942-45 US-Australian relations were much more akin to a coalition relationship – that being a temporary ad hoc arrangement, united against a specified enemy—than an alliance.[13] In the lead up to the Pacific War, Australia and the United States had very limited military-to-military contact. The two countries' armed forces were therefore largely ignorant of each other when thrust together in the face of Japanese imperialism in 1941.

While the focus of this book is on mission command, it is important to recognize that the concept as we describe it today did not exist in the period 1939-1945. This chapter will therefore address the issues of command culture and doctrine in World War

II-era Australian and US Armies in order to assess their respective approaches to command and operations. The chapter focuses on the 1943 New Guinea campaign as it was during these operations that the issues of centralized versus decentralized command were most apparent. The consideration of historical events is undertaken with our current day doctrinal definition of mission command in mind ("the practice of assigning a subordinate commander a mission without specifying how the mission is to be achieved")[14] and its implication that organizations exercising mission command conduct "military operations through decentralized execution based upon mission-type orders"[15] while judging events on the merits and doctrine of the time.

SWPA was not only the theatre in which the largest number of Australian forces operated during the war; it was also where Australians held their most senior military command positions. During earlier campaigns in the Middle East from 1940-1942, Australian operations mainly focused on the divisional level or below with only a brief foray into corps commands (Greece under General Thomas A. Blamey and Syria under Lieutenant-General John D. Lavarack). In contrast, Australian Army operations in the SWPA spanned the full spectrum from brigade to army group.[16]

The 1943 New Guinea campaign remains "the greatest concentration of Australian military power in the country's history.... Five army divisions, the majority of the RAN's [Royal Australian Navy's] capability as well as the RAAF's [Royal Australian Air Force] No.9 and No.10 operational groups, plus all attendant logistic, base, and support troops, participated in this campaign."[17] The assault on Lae and subsequent operations into the Huon peninsula highlight key issues regarding command relationships between the Australian Army and MacArthur's GHQ, the debate over the development of follow-on operations after the capture of Lae, and the command crisis during the amphibious assault at Finschhafen.

BACKGROUND: AUSTRALIAN AND UNITED STATES COMMAND CULTURE IN THE SWPA

Australia's army was fundamentally a derivative of the British Army during the Second World War. Under the Imperial Defence system reforms of 1907, the General Staff at the British War Office was redesignated the Imperial General Staff (IGS). Dominions such as Australia were told to likewise redesignate their headquarters as branches of the IGS. British and Commonwealth countries subsequently operated under common military doctrine and, largely, under common organizational tables in terms of organization of units and formations as well as equipment. The majority of senior Australian commanders during WWII were regular army officers who had been trained at British Army staff colleges at either Camberley or Quetta in India (now in modern day Pakistan). Here they were acculturated into British Army planning processes, doctrine, and methods. These officers (all major-generals and above) had also seen extensive service in the Australian Imperial Force during the First World War alongside and as a part of the British Army.[18]

Of the senior Australian officers involved in the operations in New Guinea in 1943 the Commander-in-Chief, General Sir Thomas Blamey, had seen extensive service in the First World War, had been the first Australian officer to pass the British staff college entrance examination, and was only the second Australian to graduate from a British Staff College.[19] Major-General Frank Berryman, Blamey's senior staff officer, had graduated from Camberley in 1928.[20] Blamey's senior logistics officer, Major-General John Chapman, was a 1933 graduate of Camberley. Major-General George Vasey, commander of the 7th Australian Division, graduated from Quetta in 1929.[21] Major-General George Wootten, commander of the 9th Australian Division, was a 1920 Camberley graduate, having arrived there the year earlier at the age of twenty-five already possessing extensive staff experience on the Western Front.[22] The one exception in this regard among Australia's senior ranks during the 1943 campaign was the commander of the I Australian Corps, Lieutenant-General Edmund Herring, a militia officer who thus did not attend staff college during the interwar years. Herring, however, had served in the British Army during the First World War and was a law graduate of Oxford University where he had been studying when that war broke out.

As a result of its experiences in the First World War, the British—and thus Australian—Army had adopted a decentralized approach to operations that placed a high degree of responsibility on subordinate commanders for both planning and conducting operations. The system evolved during the interwar years, the 1936 *British Field Service Regulations*, Volume III—Higher Operations noting that:

> In dealing with his subordinates, a commander will allot them definite tasks, clearly explaining his intentions, and will then allow them liberty of action in arranging the methods by which they will carry out these tasks. Undue centralisation and interference with subordinates is harmful since they are apt either to chafe at excessive control or to become afraid of taking responsibility.[23]

The guidance includes several facets of today's mission command philosophy. This said, it does not mean there was broad consensus or consistency regarding how this doctrine was applied in practice. British historian David French has argued that despite the emphasis on decentralization in doctrine, application was quite often otherwise with many commanders remaining "committed to an autocratic command and control system that inhibited subordinate commanders from exercising initiative."[24] Nonetheless, Australia's senior commanders recognized that British doctrine as it existed immediately prior to the outbreak of war emphasized a high degree of flexibility in command. This was seen to especially apply at divisional level and above where the emphasis was overwhelmingly on decentralization.[25]

Offering no little advantage in operational consistency was the presence of only one commander of the Second Australian Imperial Force (2nd AIF) from the time of its creation in 1939 through the end of the war in 1945: General Sir Thomas Blamey. Blamey was able to exert this influence even more broadly when he was made Commander-in-Chief of the Australian Military Forces (which included both the AIF and militia) in 1942. He was thereby able to set the tone for Australian command culture throughout the conflict. This stability also meant he not only directed the Australian Army's approach to command, staff work, planning, and doctrine; Blamey could in addition hand select officers he believed would most effectively support his preferred methods. While Blamey never left any specific writings outlining his views on command, planning, or staff work, history demonstrates he strongly supported the command approach articulated in the above quotation from *British Field Service Regulations*. In common with today's understanding of mission command, however, Blamey applied the approach selectively, providing different levels of detail in planning and control based on subordinate commanders' personalities and capabilities and those of their staffs.[26]

Differences in command approaches aside, Blamey recognized the importance of establishing an effective working relationship with US forces in the SWPA. The British-American Combined Chiefs of Staff had delegated command authority for the Pacific area to the US Joint Chiefs of Staff who subsequently divided responsibility into two theaters based along service lines. The US Navy under Admiral Chester Nimitz was assigned the Pacific Oceans Area while MacArthur was made Supreme Commander, SWPA. The structure therefore relied on a single joint commander for each theatre as opposed to the British model of cooperative command in a theatre through three service co-commanders. Against the direction of US Army Chief of Staff General George C. Marshall, MacArthur proceeded to fill his headquarters staff positions exclusively with US Army personnel. MacArthur's action made it clear that there would be no joint or combined (US, Australian, and Dutch) senior headquarters in the theatre. Rather, GHQ would remain exclusively American and virtually wholly US Army-dominated through the end of the war. MacArthur also made it abundantly clear that he would exercise sole joint command over all services and exercise this function at the strategic, operational, and tactical levels.[27] His approach to command combined with existing highly centralized US operational doctrine significantly retarded the practice of decentralized (mission-type) command in the Australian Army.

MacArthur's approach to his command was driven by three key factors: his antagonism towards the US Navy, his belief in the superiority of US methods over British/Australian doctrine, and his aim to maintain centralized control over operations in his theater. For the Australians this approach would be critical, as they would initially dominate the land forces component of MacArthur's command. As a result, Blamey was appointed Allied Land Forces Commander (Landops), a position he would control until late 1943. Thereafter, from the time of the assault on Lae onwards, MacArthur would effectively take command of all US ground force elements by structuring them as operation-specific task forces reporting directly to GHQ. This was done specifically to bypass the Australians, work around the US JCS directive that MacArthur could not hold the role of supreme commander and service component commander concurrently, and cement his authority at the tactical and operational as well as strategic levels of command. Despite this workaround, the critical relationships in terms of the conduct of Australian operations for the entire period of 1942-1945 would be the MacArthur-Blamey partnership and the GHQ-Landops relationship. At the center of this relationship was MacArthur's force of personality and his obsession with maintaining full control of air and naval power in his theatre.

The greatest impediment to Australian-US cooperation at the practical level was the differences in the extent of centralization practiced by the respective coalition partners. In stark contrast to the *British Field Service Regulations* quoted above, the 1939 edition of US Army Field Service Regulations 100-5, *Operations*, stipulated that "so long as a commander can exercise effective control he does not decentralize."[28] This guiding principle fitted well with MacArthur's narcissism and personal views on command and control. He sought to impose his "omnipotent notions of generalship and his assertion of total top-down control from the summit." From the time of his arrival in the SWPA, he exerted "influence in a way that probably no other individual, with the possible exception of [the US Army Chief of Staff] General George C. Marshall, could have done."[29]

Beyond differences in command philosophy, there were also fundamental disparities in the two countries' systems of planning, control, and coordination. The head of the commander's control apparatus in US Army formations was the headquarters chief of staff. This officer was foremost a staff supervisor. There was no equivalent position in the British system. Rather, the British used two senior staff officers who together were the "point of primary interaction with subordinates over the issue of orders and instructions."[30] The US system was based on the French general staff system that organized staffs into sections: G1 (personnel), G2 (intelligence), G3 (operations), and G4 (logistics). The senior general staff officer on British (and thus Australian) staffs was responsible for operations, intelligence, and training, while a senior administrative staff officer oversaw personnel and logistics. Coordination across the whole staff was carried out by the senior general staff officer when necessary.

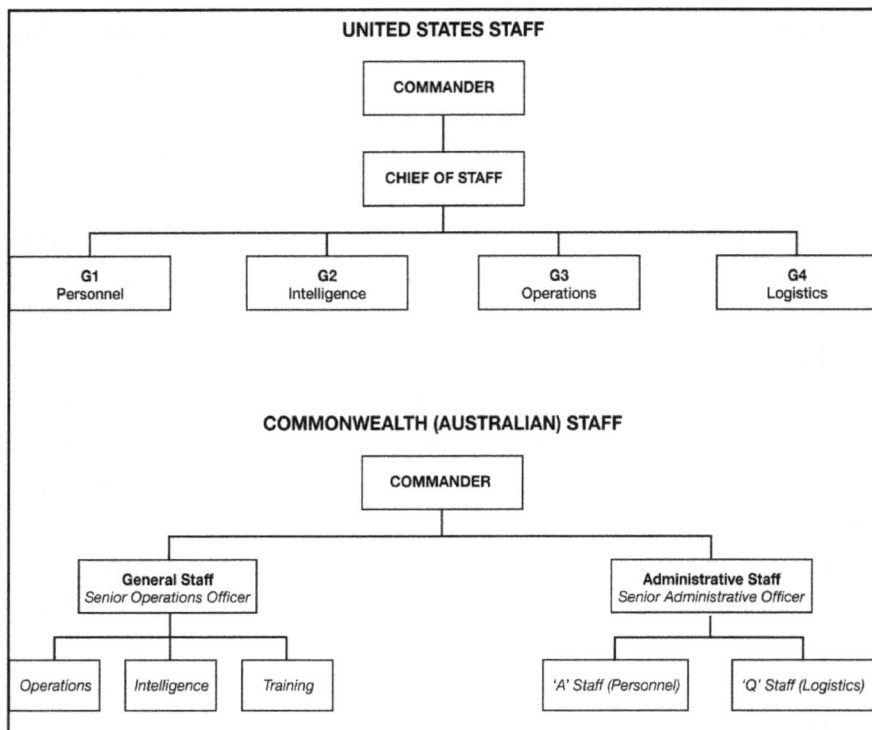

Figure 3-2: US and Australian staff systems

These differences in staff systems were not unique to the SWPA. The US staff system was employed at the senior headquarters level in both the Mediterranean and European theaters. This occurred for two key reasons. The supreme commanders in both were, at least initially, the US General Dwight D. Eisenhower. The choice was also reflective of the preponderance of US troops and resources in these theaters, as was the case in the SWPA. A third reason was that American officers found navigating the complexities of the British staff system a real challenge. Lack of previous contact between the two armies during the interwar years meant that their systems, methods, procedures, and terminology had all developed separately with no attempt to make them compatible or mutually understandable. The British staff system confused American officers in part due to its idiosyncratic divisions of responsibility that would have been recognizable to the Duke of Wellington's staff officers at the battle of Waterloo over a century before. British terminology for staff officers was also confusing, mixing as it did archaic and modern terms in a fashion that could be bewildering to the uninitiated.[31]

The US staff system was seen as an extension of the centralized managerial approach to command and control evident in US Army doctrine, an approach that was heavily emphasized in inter-war US Army education curriculums.[32] The British/Australian system was fundamentally different. A commander made decisions after which his staff converted his guidance "into orders…working out all details related to their execution [to] free his mind to deal with other, more important, matters."[33] Berryman at times found the British/Australian system very difficult to oversee as such a "decentralised command [approach required] careful co-ordination of effort [in order to] produce max[iumum] results."[34] Contrasting with the British system, MacArthur and his staff wrote SWPA plans and only after they were complete would the three service (air, sea, and land) command components become involved.[35]

Both systems had advantages. The US system produced officers notably proficient at mobilization, logistical support, and the deployment of forces.[36] This approach produced great dividends for MacArthur during the 1943-1944 period under consideration here. The Australian system was inherently more flexible. This allowed for greater responsiveness in combat as commanders were free to adapt quickly to changing operational and tactical circumstances.

MacArthur, unlike Eisenhower, not only chose not to include Australians on his staff; he gave no thought to adopting any aspects of the British command and staff system. Conversely, Blamey's Allied Land Forces headquarters was dominated by Australian officers (Blamey's request for US officers having been rejected by GHQ); he gave no thought to adopting the US staff system. The two systems therefore operated in tandem throughout SWPA operations, the major differences in organization; philosophical approaches to command; and planning and coordination of operations between GHQ and HQ, Landops continuing uninterrupted. There was to be no systematic or structural solution to these problems during the course of the Pacific War and Australian and US commanders and staffs had to find ways of working together despite the challenges. The passage of time would reflect that success in this regard became dependent on the willingness and ability of individuals to work out practical solutions, often while in contact with the enemy.

THE ASSAULT ON LAE

While Chamberlin was able to vent his frustrations to Sutherland over his August 4, 1943 exchange with Berryman, this did not put the matter to rest. As the deadline for the assault neared, American concerns regarding their lack of detailed information on the Australian plans came to boiling point. On August 14, 1943, Sutherland wrote directly to the Australian corps commander, Edmund Herring, bypassing Blamey and Berryman to demand a full detailed plan by September 1st "outlining the details for the movement of Australian and associated US forces to concentration areas as well as a detailed plan for future operations (after Lae) for the capture of Madang."[37] Chamberlin also complained to MacArthur that "the missions omitted [in the Australian plans] are more numerous than those covered…. Judged from our standard of the preparation of combat orders, it is elementary and incomplete…. It decentralises control [and] generally speaking only the initiation of the operation is complete."[38] Chamberlin had inadvertently hit on the nub of the issue: "It decentralises control"; an approach that was anathema to GHQ.

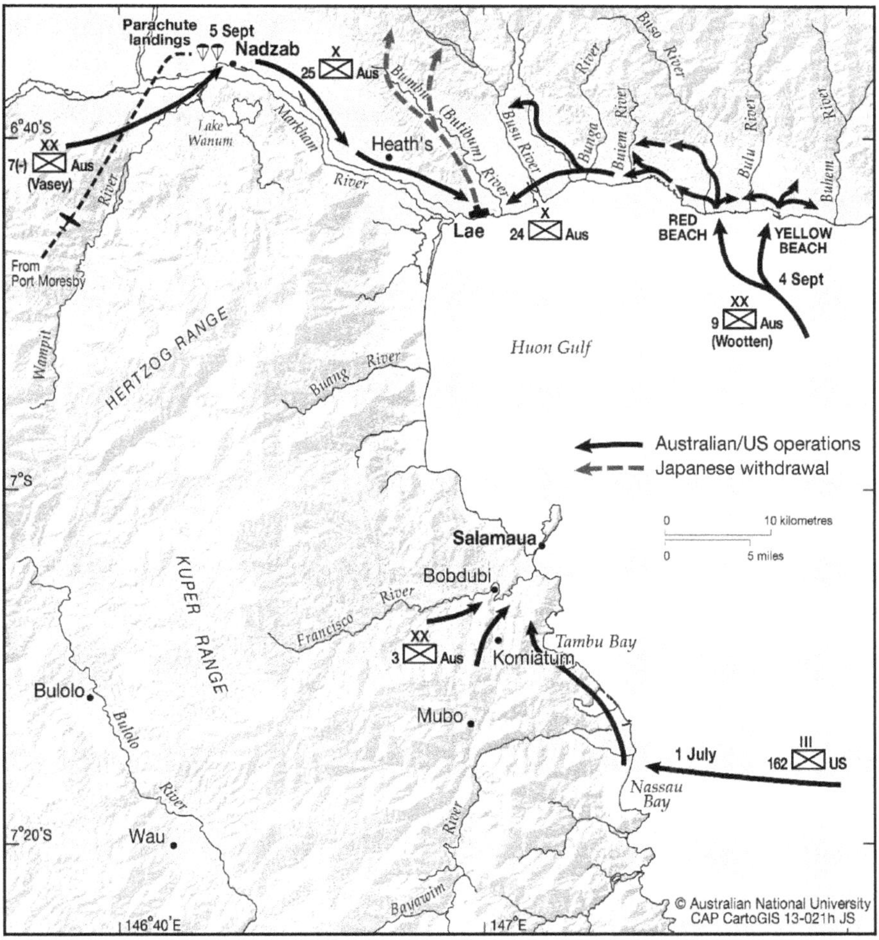

Map 3-2: 1943 Operations against Salamaua and Lae[39]

For Blamey, Berryman, and the rest of the Australians, their planning system was working as it should. Landops' role was to control high-level issues such as logistics, force concentration, and command relationships requiring coordination with GHQ. Blamey, however, also tasked Berryman and his Landops staff with additional responsibilities. They were to take over most of the initial corps-level planning, especially that pertaining to establishment of supply dumps and logistics bases and coordination of operational plans with the divisions. This was done in part to free Herring and his staff (at New Guinea Force HQ, later designated I Australian Corps HQ for the assault on Lae) to concentrate on the Salamaua deception operation, but it was also a case of Blamey selectively applying British command doctrine. Blamey believed Herring lacked the staff training and experience necessary to properly undertake such complex planning. As earlier noted, Herring had not attended staff college, nor had he any other formal staff qualification. Training of his staff was also considered insufficient, a shortcoming exacerbated by Herring's having selected officers based on close personal relationships rather than competence when manning his headquarters. Blamey therefore centralized corps planning in the hands of Berryman, who he considered the army's most experienced and expert staff planner.[40]

In contrast, Blamey and Berryman had considerable faith in the commanders of the 7th and 9th Divisions. Blamey instructed Herring to delegate tactical planning and coordination to his divisional commanders and allow them a fair degree of freedom of action. For example, the 7th Australian Division worked closely with the attached 503rd US Parachute Infantry Regiment in developing detailed plans with the US and Australian units flying them to Nadzab and otherwise providing air support for the operation. The 9th Australian Division was tasked to similarly undertake detailed planning with the US VII Amphibious Force for the assault landing west of Lae.[41]

Largely unaware of this planning and coordination as August drew to a close and D-day fast approached, GHQ remained nervous regarding Landops' ability to manage the assault on Lae. MacArthur passed on Chamberlin's concerns to Blamey and asked for a detailed response. Blamey sent MacArthur two separate notes on August 31st detailing his preparations and exchanges between GHQ, Landops staff, and subordinate unit headquarters.[42] The messages diffused the inter-headquarters tension. MacArthur,

Blamey, Sutherland, Chamberlin, and Berryman then met with the 5th Air Force commanders George Kenney and Ennis Whitehead and 7th Fleet commander Admiral Arthur Carpenter a few days later on the eve of the assault in Port Moresby to thrash out final details and discuss post-Lae operations. Broad agreement was reached on air and naval support for Australian operations in New Guinea through November 1943, support essential to capturing the Huon Peninsula. It would transpire that this was the "last time that Blamey was to be allowed to decisively influence the course of operational strategy as MacArthur's Land Forces Commander."[43]

By now the preliminary stages of Operation Postern were underway. Air superiority was the key to the success of both the amphibious landing and airborne drop. Adopting a rather different approach to command flexibility to MacArthur, US Army Lieutenant General George Kenney, Commander of the Allied Air Forces in the SWPA, quickly coordinated his US 5th Air Force with Landops and Herring's headquarters to secretly build an air base far behind enemy lines.[44] Constructed with Australian ground troop support, the base would allow Allied aircraft to bomb Japanese air bases previously beyond fighter range. The airfield was in place by July 27th and on August 17, 1943, 5th Air Force launched a major offensive against Japanese air bases in New Guinea to ensure air superiority during the Lae and Huon Peninsula operations.[45]

The assault on Lae began on September 4th with the landing of the 9th Division just beyond Japanese artillery range. The 503rd PIR secured Nadzab airfield the following day, linking up with an Australian ground force of pioneers and engineers who had marched overland behind Japanese lines to bring the facility into operation in readiness for landing the 7th Division's lead elements scheduled to arrive the following morning. Further south, the 3rd Australian Division launched its push to secure Salamaua after what had proven a highly successful deception in diverting Japanese attention from Lae.

Figure 3-3: Australian digger and American paratrooper meet for the first time near Nadzab twenty miles west of Lae[46]

The operation succeeded brilliantly. The September 4th landing was unopposed. The 503rd PIR airdrop met "practically no opposition."[47] Ten days later, the Australians captured an order for the Japanese garrison at Lae to withdraw in order to avoid encirclement. The 7th and 9th Australian divisions entered the town two days thereafter.

FINSCHHAFEN: THE CLASH OF COMMAND CULTURE

On September 17th, Berryman and Blamey attended a conference to discuss future operations with MacArthur and his senior staff at Advance GHQ.[48] Here the flexibility of the Australian approach to planning came to the fore. They had steadfastly refused GHQ's demands for a definitive plan for operations after the capture of Lae, instead developing a number of contingency plans for various extents of Japanese resistance at Lae. The plans had been developed by Landops with involvement of I Australian Corps and the 7th and 9th Division staffs.[49]

The first of these contingency plans called for major reinforcements to be sent to the I Australian Corps. These reinforcements included an armored brigade, heavy artillery, and more infantry predicated on stubborn Japanese resistance like that seen in Papua the year before. The second plan was a response to a relative swift capture of Lae and a resultantly expedited phase II of operations prior to development of Lae as a major logistical base. This alternative, strongly supported by George Kenney, involved two deep thrusts, the first by the 7th Division up the Markham Valley to secure airfields for the 5th Air Force to bomb Wewak in northern New Guinea and support operations against New Britain. The second was a coastal amphibious strike by the 9th Division against the Huon coast line at Finschhafen to secure a port as a staging base for the US 1st Marine Division's assault on New Britain and support further operations deeper into the Bismarck Sea area. With the withdrawal of the Japanese from Lae, Berryman quickly terminated preparations for the armored, artillery, and infantry reinforcements, replacing them with an Australian Base Force to develop Lae while transferring logistical priority to the two-strike option. Coordination between GHQ and Landops for these operations was undertaken from September 15-17, 1943.[50]

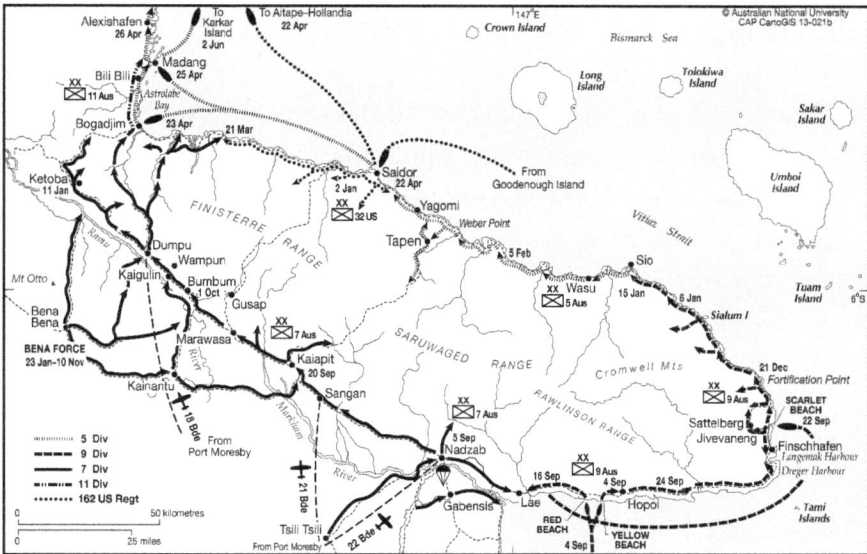

Map 3-3: Operations in and around Finschhafen, October 1943[51]

While some of the same clashes over planning would occur during development of the Finschhafen operation, the compressed timescale and the pre-existing relationship between the Australians and the flexible and adaptive approach to operations of the US 7th Amphibious Force under Admiral Daniel E. Barbey and Kenney's US 5th Air Force greatly eased tensions. Major issues would nonetheless arise as a result of MacArthur's centralized command system. While MacArthur held joint command in the SWPA, he refused to emulate this within individual areas of operation. Thus while Blamey commanded "Phosphorus," the SWPA code name for New Guinea, he did not command air or naval assets in the area but rather had to coordinate with their commanders, both of whom, like Blamey, reported directly to MacArthur. MacArthur, however, never effectively acted as a joint commander and his preference for single-service task forces meant that subordinate commanders had to bury inter-service and multinational rivalries when competing for resources to planning and executing operations.[52] Lieutenant General George Brett, commander of SWPA Allied Air Forces for much of 1942, noted:

> MacArthur is prone to make all his decisions himself, depending only upon his immediate staff. One of the perquisites of command, the coordination of the three services in a combined effort, is absolutely neglected. Commanders are not conferred with prior to either major or minor decisions. Lack of command and staff meetings results in directives impossible to interpret and orders issued without the help of those who must carry them out and who should presumably have the most specialized knowledge of the subject.[53]

This was an issue for a number of US commanders as well as the Australians. Highly successful US ground commanders in the theater; such as a Lieutenant General Robert Eichelberger

(I Corps and Eighth Army), Lieutenant General Walter Krueger (Sixth Army), and Major General William H. Rupertus (1st Marine Division—Cape Gloucester); chaffed at the excessive centralized command culture that MacArthur instigated in the SWPA.

While the situation had improved somewhat by 1943, it was only marginally better. Complicating the issue of command further: while MacArthur *de facto* controlled joint resources at all levels of command, he divided his time between the battlefront and the policy front in Australia. This system was replicated in Australia's military forces. Blamey was, in September 1943, simultaneously commander-in-chief of the Australian Military Forces (a position that included being senior military advisor to the Prime Minister and government), commander of Allied Land Forces, and commander of New Guinea Force (an operational level army command). Soon after the decision to launch the assault on Finschhafen, MacArthur returned to Brisbane to deal with matters of policy and strategy. Blamey found it necessary to follow not long thereafter. Before departing at this critical juncture, Blamey appointed Lieutenant-General Iven MacKay as interim General Officer Commanding NGF while leaving Berryman behind to act as his senior staff officer and provide continuity. MacArthur left it up to MacKay, Lieutenant General Ennis Whitehead (commander Advance Echelon, 5th Air Force, New Guinea), Admiral Daniel E. Barbey (Commander, VII Amphibious Force), and Admiral Arthur Carpenter (Commander, 7th Fleet) to cooperate with each other as operations against Finschhafen approached.[54]

This situation might have worked aside from a fundamental difference of opinion between Landops and GHQ regarding intelligence on the strength of Japanese forces on the objective, the lack of clarity in MacArthur's orders to the various commands designated to participate in the operation, and GHQ's unflagging commitment to its centralized planning process. Management issues within GHQ also continued to hinder effective coordination. GHQ tasked VII Amphibious Force to focus their planning and support on the New Britain Task Force, failing to inform the New Guinea Force that they would no longer have the amphibious support at Finschhafen after the initial landing.[55]

In terms of intelligence, GHQ insisted Japanese strength at Finschhafen was negligible, amounting to no more than 350 mostly rear echelon troops that it argued would withdraw at the first sign of an allied assault.[56] The Australians assessed that there were at least 1,000 well trained troops in the area and that the Japanese planned to use the area as a staging base to counterattack Lae. Further, any Allied operation in the area would threaten the Japanese position on the Huon Peninsula, causing them to lose control of the Vitiaz Strait between New Guinea and New Britain. The Australians therefore concluded the attack at Finschhafen would elicit a very strong response. They concluded that the Japanese were already moving new troops into the area; a Japanese infantry regiment would arrive within seven days while an additional two divisions amounting to some 20,000 troops could be in the area by October 26th.[57]

The result was a major rupture between GHQ and Landops. The former put pressure on the Australians to launch the operation as soon as possible with a minimum force. Blamey believed that the original plan to use only one brigade from the 9th Division entailed too much risk. He directed that a second brigade be added and that the divisional headquarters be sent to coordinate the operation. Blamey took the matter directly to MacArthur; after the meeting Blamey believed he had MacArthur's firm agreement that the additional forces would be allowed to move into the beachhead. GHQ, however, did not consider the additional troops necessary and held fast to its plan for transitioning priority of VII Amphibious Force support to the US Marine Corps landings at Cape Gloucester as soon as the first brigade from the 9th Division was ashore. GHQ issued its orders for the New Britain operation on September 22, 1943 with a target date of November 20th, the day of the planned assault on Finschhafen.[58]

After a rather chaotic but successful opposed landing of one 9th Division brigade at Finschhafen, Blamey ordered the follow-on forces into the bridgehead and then left for Australia. Herring, the somewhat overawed commander of I Australian Corps, attempted to coordinate the movement of the additional brigade and divisional headquarters with the VII Amphibious Force only to have his request rebuffed. VII Amphibious Force had been specifically ordered by MacArthur and GHQ to deny any such requests. What followed was a series of meetings, cables, and exchanges between the Australian Army and the US Navy in New Guinea that quickly deteriorated as each side stood firm. The issue escalated in intensity on September 26th when the Australians captured a Japanese order for a counterattack on the Australian beachhead by elements of two regiments.[59] At this point Herring's battle weariness and pessimism led him to make impertinent demands on the USN, causing a temporary breakdown in relations between the Australian Army level command, New Force Headquarters, and the 7th Fleet.

An appeal was made to GHQ; MacArthur replied, insisting the estimate of 350 Japanese troops at Finschhafen remained accurate and those troops "intend or have started to evacuate the area" despite the fact that previous Australian intelligence estimates were now reinforced by the captured Japanese order. The SWPA commander went on to decree that the only Australian "reinforcements allowed were those to keep the current troops up to strength until an airfield was built."[60]

It seems clear that MacArthur had never intended to provide any additional support from the VII Amphibious Force despite his commitment to Blamey. His refusal to recognize — or ignorance of — the tactical reality at Finschhafen and the emphasis on building an airstrip to facilitate supply and reinforcement was fanciful. Even worse, as a number of days had passed, the Australians were under attack by the Japanese and the 20th Australian Brigade in the beachhead was coming under increasing pressure to hold on. The impasse was only broken on September 28th by a direct appeal from Blamey to MacArthur. MacArthur begrudgingly conceded and ordered the navy to start moving additional forces into Finschhafen the following night.[61]

The near debacle can largely be attributed to MacArthur's defective command structure. No one aside from MacArthur—neither army commander, corps commander, amphibious force commander, nor the commander of the 7th Fleet—could order the landing of reinforcements and GHQ and MacArthur had lost touch with reality regarding the reinforcement question.[62] Japanese counter attacks were only defeated after the arrival of the additional brigade and a near disaster for the 9th Australian Division. The resultant constraints meant the Australian's had insufficient forces ashore to capture the high ground above Finschhafen immediately after landing. The Japanese were able to reinforce the area quickly and secure that key terrain. It would ultimately take four Australian brigades until November 27th to seize the Sattelberg Heights and secure the Finschhafen objective. Corps commander Edmund Herring noted soon afterward, "we damned nearly lost Finschhafen."[63]

Figure 3-4: American amphibious engineers landing a detachment of Australian troops on a north coast of New Guinea beach[64]

Lessons were learned, but many others remained to be fought over—and eventually smoothed over. The key to the success of the coalition in the SWPA mainly lies with the efforts of MacArthur's subordinates, including key staff officers and commanders from the US Army and US Navy such as Chamberlin, Kenney, Barbey, Eichelberger and their counterparts from the Australian Army such as the senior operations officer Lieutenant General Frank Berryman, senior logistics officer Major-General John Chapman, and leaders such as Lieutenant General Leslie Morshead.[65] As 1943 closed, these individuals recognized they had to find a way to make the coalition work despite different command cultures and MacArthur's omnipresent personality. However, Australia's issues increasingly took on a subsidiary role as the Americans came to dominate the SWPA through sheer force of numbers and MacArthur's personality. This overwhelming asymmetry did not mean that Australia-US friction would go away. It would again rear its head during the 1945 Bougainville and Borneo campaigns, ensuring that MacArthur and Blamey would continue to represent what would remain very different approaches to command throughout the Second World War in the Pacific.

ENDNOTES

1 L.J. Fitz-Henry 'Shells Salamaua," *Brisbane Courier-Mail* (August 4, 1943): 1.

2 Map courtesy of The Australian National University College of Asia and the Pacific CartoGIS, item no. 13-021j JS.

3 For detailed discussion of the operations in New Guinea in 1943 see Peter J. Dean, (ed.), *Australia 1943: The Liberation of New Guinea* (Melbourne: Cambridge University Press, 2014).

4 The Australians, as members of the British Commonwealth, retained command and force structure systems very similar to those of the United Kingdom during World War II.

5 Land Warfare Doctrine 1, *The Fundamentals of Land Power* (Canberra: Australian Army, 2014), 45. See details in the document's introduction.

6 Major-General Frank Berryman, diary, August 4, 1943, AWM PR84/370.

7 David Horner. *High Command: Australia and Allied Strategy 1939-1945* (Sydney: Allen & Unwin, 1982), 295.

8 Photograph from State Library Victoria, Image H30887, http://search.slv.vic.gov.au/primo_library/libweb/action/search.do?fn=search&ct=search&initialSearch=true&mode=Basic&tab=default_tab&indx=1&dum=true&srt=rank&vid=MAIN&frbg=&vl%28freeTexto%29=H30887&scp.scps=scope%3A%28PICS%29&vl%281UIStartWitho%29=contains&vl%2810247183UIo%29=any&vl%2810247183UIo%29=title&vl%2810247183UIo%29=any (accessed January 18, 2016).

9   Major General Stephen Chamberlin, letter to Lieutenant General Richard Sutherland "Status of plans for Postern," August 5, 1943, AWM 54 589/4/9.

10   Ibid.

11   Berryman, diary, August 4, 1943.

12   Chamberlin to Colonel B.Q. Jones, Commandant Joint Operational Training School, September 17, 1942, Sutherland Papers, RG-30, Box 25, Folder 8, MacArthur Archives.

13   For the prescribed definition of a coalition as adapted here see Thomas Stow Wilkins, "Analysing coalition warfare from an intra-alliance politics perspective: The Normandy Campaign 1944," *Journal of Strategic Studies* 29 (2006): 1121 – 1124.

14   The Australian Army. Land Warfare Doctrine 1 (LWD1), *The Fundamentals of Land Power*. 2014, (http://www.army.gov.au/~/media/Army/Our%20future/Publications/Key/LWD1/LWD-1_B5_190914.pdf) (accessed December 19, 2015).

15   Joint Publication 1-02, *Department of Defense Dictionary of Military and Associated Terms* (Washington, D.C.: Chairman of the Joint Chiefs of Staff, November 8, 2010 as amended through March 15, 2015), 162.

16   See Royal Australian Navy and Royal Australian Air Force official histories.

17   Peter Dean. *1943: The Liberation of New Guinea* (Cambridge University, 2013), 13; and Peter Dennis and Jeffrey Grey, eds., *The Foundation of Victory: The Pacific War, 1943-44*, Chief of Army's Military History Conference, Army History Unit, Canberra, 2003.

18 For an overview of this period see Jeffrey Grey. "Military history of Australia" and "The Australian Army," in *The Australian Centenary History of Defence: Volume I: The Australian Army* (Oxford University Press, 2001).

19 David Horner. *Blamey: The Commander in Chief* (Sydney: Allen & Unwin, 1998), 20-21.

20 Alex Hill, "Berryman, Sir Frank Horton (1894–1981)," *Australian Dictionary of Biography,* National Centre of Biography, Australian National University, http://adb.anu.edu.au/biography/berryman-sir-frank-horton-12204/text21883 (accessed September 8, 2015).

21 David Horner, "Vasey, George Alan (1895–1945)," *Australian Dictionary of Biography*, National Centre of Biography, Australian National University, http://adb.anu.edu.au/biography/vasey-george-alan-11914/text21343 (accessed September 8, 2015).

22 Alex Hill, 'Wootten, Sir George Frederick (1893–1970)', *Australian Dictionary of Biography*, National Centre of Biography, Australian National University, http://adb.anu.edu.au/biography/wootten-sir-george-frederick-12073/text21659 (accessed September 8, 2015).

23 British Army General Staff. *Field Service Regulations, Volume III: Operations- Higher Formations* (1936), 9.

24 David French. *Raising Churchill's Army: The British Army and the War against Germany 1919-1945* (New York: Oxford University Press, 2000), 13.

25 David French, "Doctrine and Organisation in the British Army, 1919-1932," *The Historical Journal*, 44 (June 2001): 515.

26   For details on Blamey's planning in the SWPA see Horner, *Blamey: The Commander in Chief* especially chapters 14-16, 18, & 22; and Dean, *The Architect of Victory,* chapters 9 -11.

27   Kevin Holzimmer, "Joint operations in the Southwest Pacific, 1943–1945," *Joint Force Quarterly* (2005), 102–3.

28   US Army War Department. FM100-5: *Tentative Field Service Regulations, Operations (1939)*, 34.

29   John Coates. "The war in New Guinea 1943–44," in *The Foundations of Victory: The Pacific War 1943-1944*, edited by Peter Dennis and Jeffrey Grey, Chief of Army's Military History Conference, Army History Unit, Canberra, 2003, 45.

30   Patrick Rose, "Allies at War: Comparing US and British Army Command Culture in the Italian Campaign, 1943-1944," (presented at the Society for Military History Conference, "Ways of War," Cantigny First Division Foundation, Lisle, IL, June 9-12, 2011), 3; and Martin van Creveld. *Fighting Power: German and US Army Performance, 1939-1945* (Westport, CT: Greenwood Press, 1982), 51.

31   Niall Barr. *Yanks and Limeys: Alliance Warfare in the Second World War* (London: Jonathan Cape, 2015), 197.

32   Eitan Shamir, "The Long and Winding Road: The US Army Managerial Approach to Command and the Adoption of Mission Command," *Journal of Strategic Studies*, 33 (2010), 649; and van Creveld, *Fighting Power,* 37.

33   John English. *The Canadian Army and the Normandy Campaign: A Study of Failure in High Command* (New York: Praeger, 1991), 89 as quoted in Rose, "Allies at War: Comparing US and British Army Command Culture in the Italian Campaign, 1943-1944," 3.

34 "Planning History and Instructions", Berryman Papers, item 48.

35 FM101-5, *Staff Officers' Field Manual: The Staff and Combat Orders*, August 1940, 1, 7. This document states "All orders from a higher to a subordinate unit are issued by the commander of the higher unit to the commander of the subordinate unit."

36 Shamir, "The Long and Winding Road: The US Army Managerial Approach to Command and the Adoption of Mission Command," 649.

37 Dean, *The Architect of Victory*, 178

38 Chamberlin to General Douglas Macarthur, as quoted in David Dexter. *The New Guinea Offensives. Australia in the War of 1939-1945: Series 1-Army (Volume VI)* (Canberra: AWM, 1961), 281; and Chamberlin to Sutherland, AWM 54 589/3/9.

39 Map courtesy of The Australian National University College of Asia and the Pacific CartoGIS, item no. 13-021h JS.

40 See Future operations – New Guinea – Report by DCGS, 11 June 1943, "Memos and Orders – Cartwheel – April–September 1943," AWM 589/3/11. See also Berryman, diary, May-August 1943.

41 Conference, Postern, HQ NGF, 25 July 1943, AWM 54 589/3/7; see also Vice Admiral Daniel E. Barbey. *MacArthur's Amphibious Navy: Seventh Amphibious Force Operations, 1943-1945* (United States Naval Institute: Annapolis, 1969) for details.

42 General Sir Thomas Blamey to General Douglas MacArthur, August 31, 1943, Berryman Papers, Item 11.

43 John Coates, "The War in New Guinea 1943-44," in Peter Dennis and Jeffery Grey, *The Foundations of Victory*, 57.

44 At this stage Lieutenant General Edmund Herring was commanding New Guinea Force, a corps level formation concentrating on the approach to Salamaua and preparing for the assault on Lae. Once Blamey arrived at Port Moresby (just before the assault), New Guinea Force would become an army level headquarters with Herring's HQ staff becoming that for I Australian Corps. I Corps formations, the 7th and 9th Australian Divisions, were the only force concentrated in New Guinea just before the assault on Lae. From this time, the 3rd Australian Division reported directly to Blamey at NGF HQ as did the Allied ground troops at Milne Bay, Wau, and other parts of Papua and New Guinea.

45 Colonel Thomas E. Griffith, Jr. *MacArthur's Airman: General George C. Kenney and the War in the Southwest Pacific. Modern War Studies Series* (University Press of Kansas, 1998), 123–8.

46 United States Army Signal Corps photograph courtesy State Library Victoria, Image H98.100/2379, http://search.slv.vic.gov.au/primo_library/libweb/action/search.do?fn=search&ct=search&initialSearch=true&mode=Basic&tab=default_tab&indx=1&dum=true&srt=rank&vid=MAIN&frbg=&vl%28freeText0%29=h98.100%2F2379&scp.scps=scope%3A%28PICS%29&vl%281UIStartWith0%29=contains&vl%2810247183UI0%29=any&vl%2810247183UI0%29=title&vl%2810247183UI0%29=any (accessed January 18, 2016).

47 Berryman, diary, September 4-5, 1943.

48 Notes for Memorandum, Conference G-3, GHQ, "Operations Markham-Ramu and Vitiaz Strait Areas," RG407 GHQ SWPA Operational Reports, 98-GHQ1-3.2 G-3 Journal and & Files

Box 598, September 1943, NARA, Maryland, VA.

49 Dean, *The Architect of Victory*, 230, 249; and Garth Pratten, "Applying the Principles of War: Securing the Huon Peninsula" in Peter J. Dean, (ed.), *Australia 1943: The Liberation of New Guinea*, 257.

50 Pratten, "Applying the Principles of War: Securing the Huon Peninsula," 257.

51 Map courtesy of The Australian National University College of Asia and the Pacific CartoGIS, item no. 13-021b.

52 Holzimmer, "Joint operations in the Southwest Pacific, 1943–1945," 102–3.

53 Lieutenant General George Brett to General George Kenney, Kenney Diaries, Volume I, RG54 MacArthur Memorial Archives, Norfolk, Virginia.

54 See Dean, *The Architect of Victory*, 249-255.

55 Horner, *High Command*, p. 300.

56 Coates, "The war in New Guinea 1943-44," 64.

57 'Operations for the Capture of Madang', Berryman Papers, item 37.

58 John Miller, Jr. *The United States Army in World War II, The War in the Pacific, Cartwheel: The Reduction of Rabaul* (Washington: Officer of the Chief of Military History, 1959), 272–3.

59 Dexter, *The New Guinea Offensives*, 479.

60 New Guinea Force to I Australian Corps, September 28, 1943,

reply in quote of GHQ signal, AMW 54 591/7/21.

61   Horner, *Blamey: The Commander in Chief*, 424.

62   Pratten, "Applying the Principles of War," 266.

63   Edmund Herring as quoted in Horner, *Blamey*, 424.

64   United States Army Signal Corps photograph courtesy State Library Victoria, Image H98.100/2252, http://search.slv.vic.gov.au/primo_library/libweb/action/search.do:jsessionid=918E1D5EFE6F732193C1ABE41EB55907?fn=search&ct=search&initialSearch=true&mode=Basic&tab=default_tab&indx=1&dum=true&srt=rank&vid=MAIN&frbg=&vl%28freeText0%29=H98.100%2F2252&scp.scps=scope%3A%28PICS%29&vl%281UIStartWith0%29=contains&vl%2810247183UI0%29=any&vl%2810247183UI0%29=title&vl%2810247183UI0%29=any (accessed January 18, 2016).

65   For a detailed analysis of the coalition relationship on the battlefield, see Peter J. Dean, *MacArthur's Coalition: US and Australian Military Operations in the Southwest Pacific Area, 1942-1954* (Lawrence: University Press of Kansas, 2018).

4

THE TIES THAT BIND: AUSTRALIA, MISSION COMMAND, AND THE KOREAN WAR (1950-1953)

Dr. Meghan Fitzpatrick

INTRODUCTION

The Australian Defence Force's (ADF's) inclusion of mission command as part of official doctrine represents a relatively recent development.[1] Yet the idea is far from new. Australian doctrine has long emphasized the need for officers at all levels to seize the initiative when the opportunity presents itself. The Australian soldier acquired a reputation for aggression and resourcefulness on WWI and WWII battlefields.[2] Operations during the Korean War (1950-1953) served to further enhance this image.

The Korean War was a brutal conflict that pitted North Korean and Chinese forces against a United Nations (UN) coalition headed by the United States. Described by American military researcher SLA Marshall as the century's "nastiest little war," the fighting in Korea produced roughly four million casualties in the space of three years.[3] It was the first time that the Cold War turned hot and a crucial test of the UN as an organ of international security. Over 17,000 Australians fought in the conflict, one of the largest deployments of Australian troops in history.[4] This chapter explores the Australian understanding of the mission command concept during the Korean conflict and how successful the contingent was in employing this philosophy within the framework of a larger Commonwealth and United Nations alliance. The chapter concludes by considering the lessons that the Korean War experience provides for multinational operations.

ORIGINS OF AUSTRALIAN PARTICIPATION

The Korean War first erupted in the early hours of June 25, 1950 when communist North Korea launched an invasion of South Korea in a bid to unify the two countries. While the attack was largely unforeseen, the UN responded quickly and called on its member states to "furnish such assistance to the Republic of Korea [South Korea] as may be necessary to repel the armed attack and restore international peace and security in the area."[5] Shortly thereafter, the Americans sent a battalion of men based in Japan to Korea. US Army General Douglas MacArthur was appointed as the Commander-in-Chief, United Nations Command (UNC).[6]

Alongside other Commonwealth countries, Australia was quick to respond to the UN's call for help. The destroyer Her Majesty's Australian Ship (HMAS) *Bataan* and the frigate HMAS *Sholhaven* were en route to Korean waters within days. The United Kingdom, Canada, and New Zealand followed suit by contributing naval forces in the weeks following.[7] American forces experienced a series of major setbacks during the summer of 1950 and were compelled to withdraw to a perimeter around the city of Busan on Korea's south coast. The US Army had suffered a staggering 15,000 casualties by early August.[8] In consequence, Commonwealth countries came under substantial pressure to commit ground troops to the campaign.

The Australians in particular had a vested interest in the success of UN forces in Korea. Prime Minister Robert Menzies was a determined opponent of communism.[9] Moreover, Minister for External Affairs Percy Spender saw Korea as a critical opportunity to cement Australia's relationship with the United States. In the summer of 1950, Australia and New Zealand were still in the process of negotiating what would become the landmark Pacific Security or ANZUS Treaty. Spender immediately recognized that deploying ground troops to Korea would help secure one of Australia's key foreign policy objectives.[10] Soldiers of the 3rd Battalion Royal Australian Regiment (3 RAR) received notice that they were bound for Korea on August 2, 1950.[11]

Landing at Busan on September 28, 1950, 3 RAR was under-strength and largely unprepared for war. Deploying from duties with the British Commonwealth Occupation Force in Japan, the battalion had only weeks before been well below its wartime establishment of 960 with only twenty officers and 530 other ranks.[12] Under Australian law, members of the battalion were required to volunteer for service. While all but twenty-six signed up, further reinforcement was need-ed. Volunteers from the 1st and 2nd Battalions of the Royal Australian Regiment (1 RAR and 2 RAR) were transferred to 3 RAR to bring the unit up to strength.[13] The battalion was initially filled with soldiers who were unfamiliar with each other and their leaders. Over time, this was remedied by an intense schedule of training.

Be that as it may, the Americans were in urgent need of help and the unit immediately began to prepare for deployment.[14] In addition, the government announced the recruitment of a special force that provided a number of reinforcements to 3 RAR as it readied to deploy. The members of this K Force had to be between the ages of twenty and thirty-nine and with previous armed forces service of at least three years. The vast majority of K Force volunteers were battle-hardened veterans of WWII.[15]

Upon landing in Korea, 3 RAR was assigned to the 27th British Commonwealth Brigade commanded by British Army Brigadier Basil Aubrey Coad alongside the 1st Battalion of the Argyll and

Sutherland Highlanders and 1st Battalion of the Middlesex Regiment. The brigade was part of the US IX Corps. 3 RAR would remain in Korea for the rest of the war.[16]

Arriving shortly after the American landings at Incheon, 3 RAR joined the campaign as UN forces scored a series of major victories and advanced up the peninsula towards the Sino-Korean border.[17] It appeared as if the Australians would see little fighting in that early autumn of 1950. However, events quickly took a turn for the worse. Chinese troops, dispatched by authorities in Beijing who feared the war would expand into Manchuria, began to cross the border in late October. There were roughly 250,000 Chinese soldiers in theater by mid-November and their numbers increased daily.[18] UN forces were forced to withdraw southwards.

Fighting remained highly mobile and volatile until the summer of 1951, the front line settling along the 38th parallel as peace negotiations began. This new phase of the war coincided with the creation of the 1st British Commonwealth Division, part of US I Corps, and in turn the American Eighth Army. Commanded by the British Army's Major-General Jim Cassels, the division was originally composed of the 28th British Commonwealth Brigade, 25th Canadian Infantry Brigade, and 29th British Infantry Brigade. The Australians were assigned to the 28th Brigade that had relieved the 27th Brigade several months earlier in April.[19]

The Commonwealth and Mission Command

Australian leaders exercised a form of control that we would today recognize as mission command during both the mobile and static phases of the war. Historian Robert O'Neill noted that the Korean War would show "the Australian soldier had lost none of the versatility, toughness, and initiative which were the hallmarks of his predecessors in the First Australian Imperial Force (AIF) and the Second AIF."[20]

The Australians were members of the Commonwealth Division throughout the war. Established in July 1951, the division was formed largely because each contingent lacked the necessary manpower and resources to function independently. Like all marriages of convenience, disagreements were common and friction inevitable. Increasingly confident and independent, the old dominions wished to be recognized as the United Kingdom's equal partner. The British were not always sensitive to that consideration, much to their own detriment. Others like the Canadians, for example, had resisted joining the division for over a year and only relented in the face of pressure from Washington.[21] Even after July 1951, they liked to keep to themselves and sometimes employed tactics that riled their allies.[22] Despite these tensions, the members of the division shared far more similarities than differences. The Commonwealth countries largely managed to cooperate effectively in the field. Most importantly, key elements that underpin what we would today call mission command were present, namely mutual trust and good communication. A correspondent for the divisional broadsheet *Crown News* commented following the armistice, "more important than our success in battle, we have learnt that we can rely on one another. To all of us who have had the honour of serving in the Division, the Commonwealth is no longer an abstract and vague idea culled from the text books and news papers, but a reality."[23]

The Commonwealth countries had a proud history of working together on the battlefield. They had fought side by side in theaters around the globe during WWI and WWII. In addition, their militaries had a longstanding tradition of sending officers to train at British military schools and vice versa, potentially a notable consideration as officers sent to Korea were selected for their experience and ability to work effectively within a coalition environment.[24] There were also strong similarities between Commonwealth forces that further enhanced their ability to work together. For example, in the early 1950s Australian and British forces were similarly structured, used the same equipment, and employed common operational terminology. Moreover, they shared combat doctrine.[25] They likewise critically shared a philosophy of developing officers comfortable with delegating authority downwards.[26] Chief of the Imperial General Staff Field Marshal William Slim best summarized the Commonwealth perspective in his seminal 1956 book *Defeat Into Victory*. Slim argued that by providing officers with latitude to determine their own plans, they in time "developed a...flexibility of mind and a firmness of decision that enabled them to act swiftly," concluding, "this acting without orders, in anticipation of orders, or without waiting for approval, yet always with the overall intention, must become second nature and must go down to the smallest units."[27]

Mission Command at Maryang San

The Battle of Maryang San (October 3-7, 1951) exemplifies Commonwealth, and, in particular, Australian, practice of mission command in Korea. Described as a "classic battalion operation," Maryang San has also been characterized by Australia's official Korean War historian as the "greatest single feat of the Australian Army during the Korean War."[28] Reflecting on events, a veteran later recalled that the "teamwork displayed in 3 RAR...by individual soldiers and officers was superb—at all the team levels from the unit down to the two forward scouts."[29] 3 RAR managed to oust a numerically superior force from a well-defended position during five days of fierce fighting.[30]

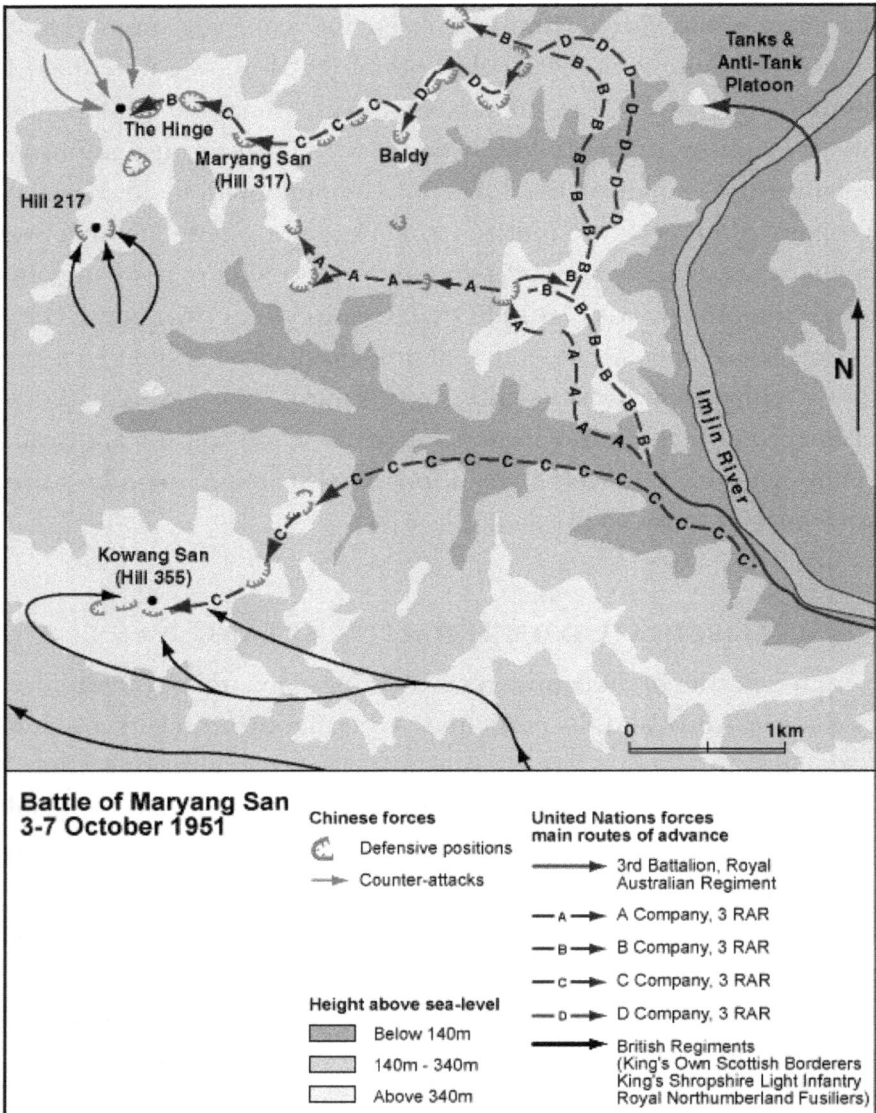

Map 4-1: Mission command in practice—The Battle of Maryang San[31]

By early October 1951, peace negotiations between UN representatives and their Chinese and North Korean opposites had fizzled. Hoping to improve the UN's position at the negotiating table, the Commander-in-Chief UNC decided to launch a limited offensive against enemy forces positioned along the Imjin River.[32] As part of I Corps, the Commonwealth Division was expected to advance from 6,000 to 8,000 yards to seize a number of critical defensive positions from the Chinese.[33] 28th Brigade was assigned the task of capturing two key features: Hill 355 (Kowang San) and Hill 317 (Maryang San). The King's Own Scottish Borderers (Kosbies) were to capture Hill 355 with the support of the King's Shropshire Light Infantry and 3 RAR in the operation's first phase.[34] 3 RAR was to secure "hill 317...a steep, bare, and conical shaped hill rising right out of the valley on its south-east approach" in phase II.[35] This was an ambitious plan involving a "long approach march with open flanks and a penetration of about three kilometers through strongly held...positions where the Chinese were well armed, well-equipped and determined to stay."[36] Rugged terrain and thick woods surrounding the hill further complicated the task.[37]

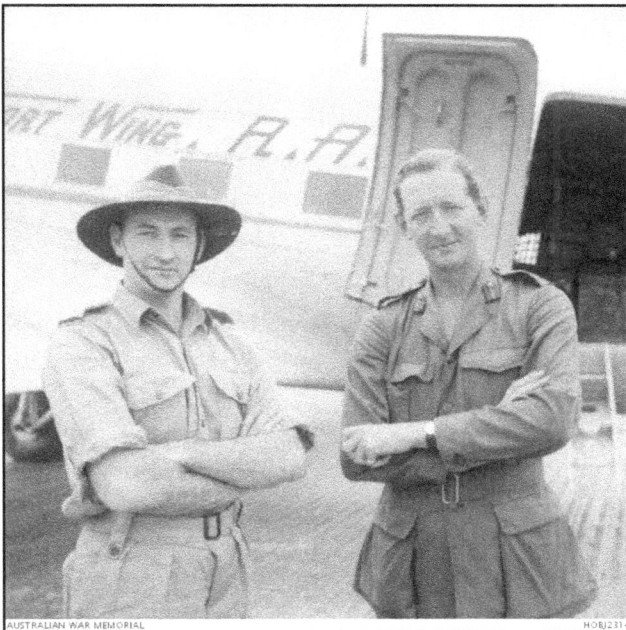

Figure 4-1: Lieutenant Colonel Hassett (left) poses for a photograph with Major General Cassels prior to taking over his new post as commanding officer of 3 RAR in July 1951.[38]

Having assumed command of 3 RAR in July, Lt. Col. Francis G. Hassett would play a pivotal role in the success of the early October 1951 battle for Maryang San. A Duntroon graduate, he had served in North Africa and Papua New Guinea from 1939 to 1945, rising to become the Australian Army's youngest lieutenant colonel.[39] Hassett was initially concerned regarding the risks involved in executing the plan when he was informed of 3 RAR's role in Operation Commando. In a 2001 article for the *Australian Defence Force Journal*, he argued that "British commanders still look[ed] on Australians as 'shock troops' and [were] not averse to using them as such."[40] Hassett and others within the battalion believed that the unit had been "pushed too hard [and] shoved into danger spots," during the first year of the war.[41] However, Hassett's anxieties were greatly alleviated by the brigade's commander, Brigadier George Taylor. Born in 1905, Taylor had first joined the British Army in 1929 as an officer in the West Yorkshire Regiment. He served in Northwest Europe throughout WWII and acquired a reputation for "coolness under extreme stress" during the Normandy campaign.[42] A decorated officer, he was put in command of the 28th Brigade in the spring of 1951. In a 1994 interview with the Australian War Memorial, Hassett recalled Taylor as an experienced soldier and a very good tactician.[43]

Hassett also had a healthy respect for the skills and abilities of the division's first General Officer Commanding (GOC), Major General Jim Cassels. Commissioned into the Seaforth Highlanders in 1926, Cassels had served in many different theatres by the time he was dispatched to Korea. Like Hassett, he had risen through the ranks rapidly during WWII to become the "youngest divisional commander in the British Army."[44] He had previously served as head of the UK Services Liaison Staff in Canberra and forged strong bonds with his Australian colleagues.[45] As noted historian Jeffrey Grey has pointed out "a man better qualified to head the composite Commonwealth Division and deal with potentially troublesome dominion forces would have been hard to find."[46] Hassett later characterized Cassels as a "very impressive man" who had a "great touch with soldiers at all levels."[47]

Taylor designed his brigade plan in such a way as to allow "junior officers to add their initiative to it as events flowed."[48] He collaborated closely with Hassett, knowing he was well versed in infantry tactics given the expertise established over sixteen years in the military.[49] Hassett refined the brigade guidance for the battalion by carefully studying the area surrounding Maryang San from the air and on the ground as he prepared the 3 RAR plan. Moreover, he conferred exhaustively with the US battalion commander whose men would advance on 3 RAR's right flank. The Americans had attacked Hill 317 on several prior occasions but failed to secure the objective. Hassett believed that they had failed because they had attacked during the day when they could easily be pinned down by devastating crossfire from heavily fortified Chinese positions. Consulting with his rifle company commanders, he soon concluded that they could exploit the element of surprise by advancing under the cover of darkness.[50] He further observed that "the technique of 'running the ridges' which had been used successfully by the Australians against the Japanese in New Guinea utilizing the tactical superiority of high ground, the cover afforded by the jungle and the relative ease of movement along a crest-line" would work with Maryang San.[51] His men would advance along its ridges to avoid the full ferocity of the enemy's firepower rather than attack directly up the slopes of Hill 317.

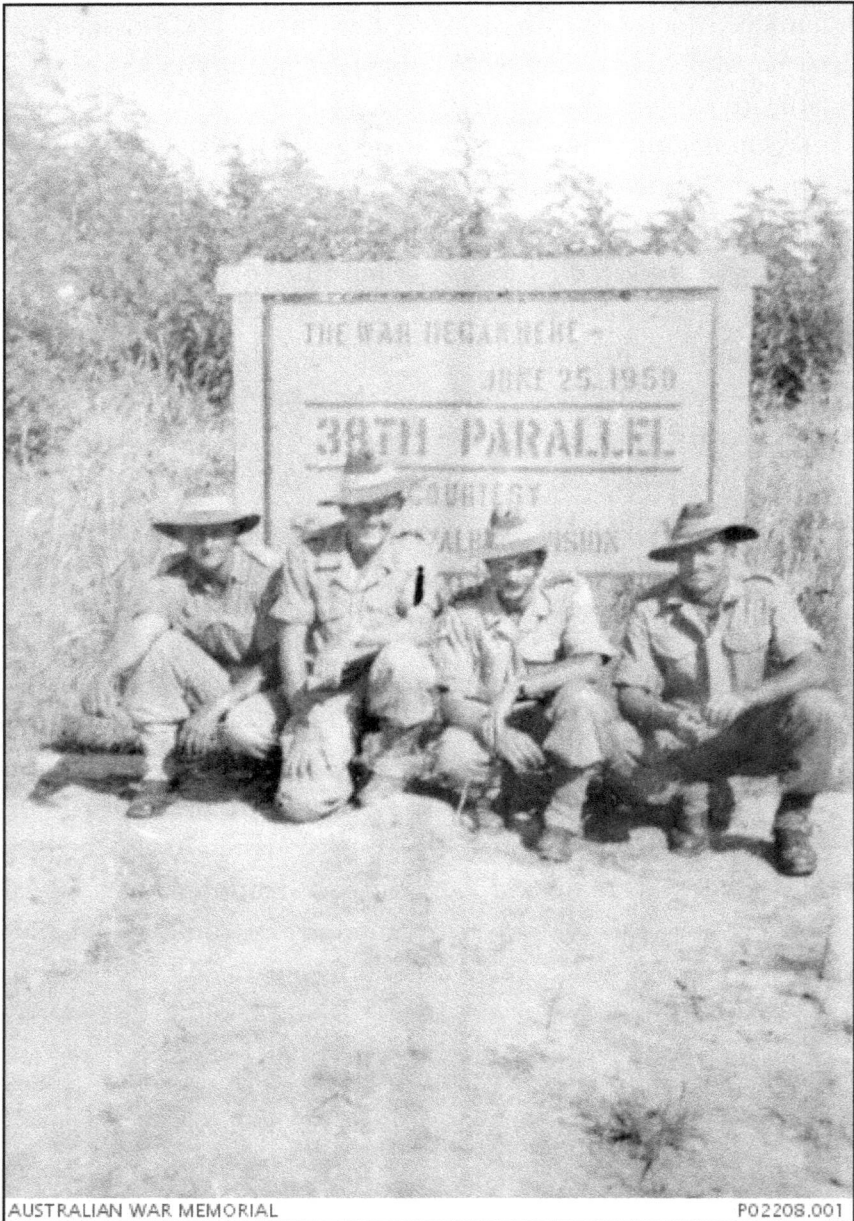

AUSTRALIAN WAR MEMORIAL   P02208.001

Figure 4-2: A portrait of the 3rd Battalion Royal Australian Regiment's four rifle company commanders. From left to right: Major Jack Gerke (C Company), Major Jeffrey James Shelton (A Company), Captain Henry William Nicholls (B Company), and Major Basil Hardiman (D Company) in September 1951.[52]

When asked to reflect on the lessons of Maryang San several decades after the war, Hassett was unstinting in praise of his four rifle company commanders. He stressed that "a battalion commander must select his company commanders carefully. He cannot risk a weak officer in these key appointments. The company commander is about the highest-level of close, direct personal contact with soldiers. It is a big job for a young man, responsible at times for the lives of a large group of soldiers."[53] In October 1951, Hassett had followed his own advice selecting his rifle company commanders with exceptional care. They included Major James Shelton (A Company), Major Henry William Nicholls (B Company), Major Jack Gerke (C Company), and Major Basil Hardiman (D Company).[54] Both Nicholls and Gerke were decorated WWII veterans. Like Hassett, they had served in New Guinea and were familiar with fighting in mountainous terrain. Hassett chose B Company to lead the advance on Hill 317 primarily because of the depth of Nicholls' experience and his infectious enthusiasm when leading men into battle. Gerke was put in charge of the reserve company. Explaining his rationale to an interviewer in 1994, Hassett argued that the reserve was likely to be assigned tasks suddenly and there was therefore often little opportunity to discuss how they should be carried out. Hassett felt that Gerke—an experienced company commander who was known for efficiency—was the best choice to shoulder this responsibility.[55] Like his compatriots, Hardiman had also seen service from 1939 to 1945; he was described as "hard-driving and brave" one who had carried out a previous raid on Hill 317.[56] In contrast to his colleagues, Shelton had only graduated from Duntroon in 1946; Korea was his first operational posting. Yet he had served in Japan for several years and acquitted himself well since joining 3 RAR in April. He was chosen to lead A Company up Maryang San's southeast slopes as a diversion.[57]

Phase I of Operation Commando began before first light on the morning of October 3, 1951. A and B Companies set off through the morning mist, following the King's Own Scottish Borderers. The Kosbies failed to secure Kowang San by evening. Instructing them to dig in along the lower slopes of the hill, Taylor ordered Hassett to renew the assault. Major Gerke's C Company launched a surprise attack in the early morning hours of October 4th. Gerke and his men managed to clear the hill of Chinese troops after brutal fighting. After securing the objective, they were instructed to hold their position until relieved by the Kosbies.[58]

Phase II of the operation commenced on October 5, 1951. With surprise lost, Hassett knew that Maryang San would be tough to capture. He ordered A Company to approach the objective along a ridgeline from the east to distract the enemy while B and D Companies attacked along a spur from the northeast. C Company remained in reserve. B and D Companies were close to securing their objectives by 0930 hours when the dense fog that had been obscuring their advance suddenly lifted.[59] Exposed to withering fire, D Company was soon engaged in a ferocious close quarter battle. Major Hardiman guided the men forward, continuing despite heavy bleeding from a thigh wound after being wounded by machine gun fire. Platoon commander Lieutenant Jim Young took control of D Company following Hardiman's evacuation.[60] Reassuring Hassett over the radio that the unit could continue the attack, Young requested twenty minutes of heavy supporting fire and renewed the advance.[61]

Having suffered significant casualties, D Company struggled to maintain the momentum of the attack. Ordering Young to dig in, Hassett called on Gerke and C Company for the second time in two days. Gerke and his men secured their objective, 3 RAR continuing to battle through the next day to clear surrounding features of remaining Chinese defenders.[62] The Kosbies relieved 3 RAR after five days of fighting, allowing the battered unit to withdraw.

Hassett credited 3 RAR's outstanding success at Maryang San to teamwork, initiative, and leadership. While he readily acknowledged the bravery of his Chinese opponents, Hassett argued that they were "less effective because of lack of initiative or restraints imposed at the various levels of command."[63] In contrast, the Australian subordinate commanders had been allowed to take calculated risks. "Initiative played a big part in winning the... battle," Hassett observed; "there were times when issues hung in the balance. The lead taken at such times by individuals regardless of rank, swayed the issue."[64]

The Battle of Maryang San provides a compelling example of mission command at work from divisional commander to platoon commander level. Senior commanders granted junior officers the opportunity to exercise their judgement and skill at pivotal moments. Trust underlay the freedom of action granted by commanders at the division, brigade, and battalion levels. Proven performance served as the foundation for that trust as it did for the selection of the leaders put in key positions. Moreover, these leaders were well briefed regarding mission objectives and the challenges they would face. Company Sergeant Major Arthur Stanley described the taking of Kowang San and Maryang San as "one of the best planned operations" in which he ever took part.[65] Excellent leadership complemented the exercise of mission command. Speaking with historian Bob Breen, Jack Gerke explained that Hassett "constantly moved amongst the men and kept us well informed. He was an inspiration to every man in the battalion and gave us confidence."[66] Aware of the capabilities and limitations of those under his command, Hassett trusted his officers to rise to the occasion.

Shelton and Nicholls were awarded the Military Cross. Gerke received the Distinguished Service Order (DSO). Hardiman was mentioned in dispatches.[67] Hassett's British superiors accorded him a similar level of recognition by awarding him the DSO. Brigadier Taylor had nothing but praise for Hassett, who he saw as an "outstanding commanding officer" with exceptional "tactical skill."[68]

Operation Commando represented the last significant advance by UN forces during the Korean War. Attrition replaced mobility during the ensuing months, operations being typified by "night-patrols in no-man's-land and set piece offensives launched from trenches across enemy minefields and barbed wire with massive artillery support."[69] Australian battalions dominated no-man's-land, a success largely attributable to a rigorous schedule of patrolling and training. For example, 1 RAR sent out twenty-eight ambush and twenty-seven fighting patrols in the first three weeks of September 1952 alone.[70] Australian battalions were sending an average of sixty-five men out on patrol every night by early 1953.[71] Their methods were considered so effective that they were the subject of a 28th Brigade memorandum distributed throughout the Commonwealth Division. The memo highlighted the importance of aggression, initiative, and creativity. Similar guidelines underlined the role of company and platoon level commanders in determining the nature and extent of their patrols.[72]

Australians trained intensively and focused on the development of junior officers when in reserve. Here as in battle, senior commander intentions were apparent in instructions. Company commanders were provided with a list of topics and directed to develop training programs after the battalion was relieved in late November 1952. Theirs was the authority to design a syllabus based on personal appraisals of the battlefield and their unit status.[73]

Anglo-Australian relations were further enhanced with the June 1952 appointment of Brigadier TJ Daly as commander 28th Brigade. A 1933 Duntroon graduate, Daly had a prestigious record of service including the command of the 2/10th battalion during the Balikpapan landings in 1945 Borneo. Appointment of an Australian officer to brigade command reflected both the increasingly Australian composition of the brigade and the respect in which the contingent was held. Daly's successor was fellow countryman Brigadier JGN Wilton.[74]

THE UNC AND MISSION COMMAND

As part of the Commonwealth, the Australians were insulated from the higher UN command authority in a manner that would not have been the case if they had been directly attached to an American formation. While the US Army had adopted aspects of mission command as early as 1939, they continued to exercise what has been described by Israeli scholar Eitan Shamir as a "managerial approach" to leadership.[75] He contends that the "managerial approach," shaped by business management theory, is distinguished by "centralization, standardization, detailed planning, quantitative analysis and aspires for maximum efficiency and certainty."[76] This style of command is also characterized by a thirst for information at the top and close control of subordinates.[77] Contrasting sharply with the philosophy of mission command, there were unsurprisingly marked differences between the Commonwealth and American command approaches in Korea. American officers also had political reasons to exercise such tight control and employ a more prescriptive approach to leadership, especially during the static phase of the war. After all, any movement along the frontline could adversely impact the progress of negotiations with the North Koreans and Chinese and further delay the signing of an armistice.

These distinctions remained a constant source of friction throughout the war. American officers had a tendency to micromanage and interfere with the activities of their subordinates as the Korean campaign evolved into a war of attrition. In his book *War of Patrols*, Canadian author William Johnston is quick to point out that there "can be little doubt that as the war in Korea stalemated, higher American commanders injected themselves to too great a degree into small unit operations."[78] Throughout his tenure as GOC of the Commonwealth Division, Maj. Gen. Cassels frequently butted heads with the commander of I Corps, Lieutenant General John W. O'Daniel. Privately, Cassels described his superior as a "Two Gun Patton" type who was willing to "undertake foolhardy stunts" for little benefit.[79] Upon his appointment as GOC, Cassels was issued with a directive that empowered him to refuse any orders that he believed would unduly endanger his command. Cassels considered using his directive on five separate occasions in the first three months of his tour.[80]

The division's second GOC, Major General Mike West, shared an equally tense relationship with O'Daniel's successor, Lieutenant-General Paul W. Kendall. By early 1953 there were also "signs of increasing centralisation" of command.[81] West noted in one of his periodic reports that "strict, almost hide-bound, orders are in force regarding the briefing of patrols and any setback, however minor, is the subject of a searching inquiry in US units — indeed the unfortunate patrol commander may even have to report to the army commander."[82] This tension eventually culminated in Kendall's dismissal in April 1953 when he "publicly rebuked West" and made "disparaging remarks about the British" at a meeting of divisional commanders.[83]

Commonwealth and American officers took a different approach to everything from patrolling to the control of artillery fire. Consequently, the relationship between the division and its senior partner was characterized by a sense of mistrust. Elements that supported the practice of mission command in the Commonwealth units were noticeably absent within the wider framework of the UNC. The relationship was further hampered by confused and poor communication. Commonwealth officers complained that US Army orders were couched in vague language.[84] As subordinate officers were given latitude on determining how best to execute their orders within the Commonwealth Division, the commander's overall intent needed to be defined as clearly as possible. The division eventually began to translate all of the orders issued by I Corps headquarters into Commonwealth parlance before they were passed on to the brigades. The division also issued a lengthy memo explaining the differences in how

Commonwealth and American forces were structured for the benefit of their US Army counterparts.[85]

Concluding Observations

The Australian battalions clearly exercised a form of mission command during the Korean War. Developing a formidable reputation for skill, Australian officers were encouraged to act independently and aggressively in the field. The British, Canadians, Australians and New Zealanders all shared a common military tradition and approach to the practice of leadership. As Field Marshal Cassels later explained in a 1994 interview, "it was usual to pass detailed orders regarding objectives and formation boundaries down the chain of command" within the Commonwealth system "but to leave the method of carrying out these orders to the discretion of the subordinate formation commander."[86] However, the Australians and their allies were constrained by the nature of the war. While a high tempo of maneuver during conventional war encourages the practice of mission command, operations in Korea closely resembled those conducted on the Western Front during WWI from the autumn of 1951. Korea was a limited war in which the UNC only employed force as a means to improve its position at the negotiating table in its latter months. Consequently, "the conduct of operations was characterised by detailed planning and preparation." The employment of "firepower and initiative was limited to that movement which could be carried out within range of direct and indirect supporting fires."[87] Commonwealth operations were also constrained by the American managerial style of command, one in which "subordinate commanders received not only their objectives but also a set of instructions as to how to achieve them."[88]

For those who have served in Iraq and Afghanistan, the Australian experience in Korea underlines an oft repeated and all-too-familiar lesson: working with allies is inherently difficult. This is especially true if the partners in question do not share the same operational terminology or a common approach to matters of doctrine. Major differences will inevitably result in disunity, friction, and possibly to resentment. Mission command relies heavily on trust and communication. Close allies can overcome their differences by fostering a relationship during periods of peace through joint exercises, exchange of officers, and common military education. But trust does not develop overnight. It is a product of time and patience.

ENDNOTES

1   Australian Defence Doctrine Publication 00.1: Command and
    Control (Canberra: Department of Defence, 2009), 14.

2   B.M. Young, "Application of Brudenell White's Command and
    Staff Style to the Modern Australian Army," *Australian Defence
    Force Journal (ADFJ)* No. 111 (March/April 1995): 11-16.

3   S.L.A. Marshall as quoted in Dan Bjarnason, *Triumph at
    Kapyong: Canada's Pivotal Battle in Korea* (Toronto: Dundurn,
    2011), 11; Richard Whelan, *Drawing the Line: The Korean War,
    1950-1953* (London: Faber & Faber, 1990), 373; Max Hastings,
    The Korean War (UK: Macmillan, 2000), 7, 45-48, 438; Herbert
    F. Wood, *Strange Battleground: The Operations in Korea and their
    Effects on the Defence Policy of Canada* (Ottawa: Queen's Printer
    and Controller of Stationary, 1966), 10-12, 42; and Tim Carew,
    The Commonwealth at War (London: Cassell, 1967), 13, 27, 71.

4   Richard Trembeth, *A Different Sort of War: Australians in Korea
    1950-53* (Melbourne: Australian Scholarly Publishing, 2005), 1.

5   UN Security Council Resolution 83 (1950), June 27, 1950
    [S/1511] as quoted in Evan Luard, *A History of the United
    Nations: The Years of Western Domination 1945-55*, Vol I (London:
    Macmillan, 1982), 241-242.

6   David Smurthwaite and Linda Washington, *Project Korea:
    The British soldier in Korea, 1950-1953* (London: National Army
    Museum, 1988), 14; Hastings, *Korean War*, 7, 438; and Carew,
    Commonwealth at War, 13, 27.

7   Anthony Farrar-Hockley, *The British Part in Korean War*, Vol. II,
    An Honourable Discharge (London: HMSO, 1995), 295-329; and
    "War at Sea: the Royal Australian Navy in Korea," Australian

War Memorial (AWM), http://www.awm.gov.au/exhibitions/ korea/ausinkorea/navy/ (accessed August 19, 2015)

8   Carew, *Commonwealth at War*, 27, 71; Fairlie-Wood, *Strange Battleground*, 42; and Hastings, *Korean War*, 96-115.

9   Cameron Forbes, *The Korean War: Australia in the Giants' Playground* (Australia: Macmillan, 2010), Kindle Edition, loc. 329.

10   Ian McGibbon, "Anzaxis at War: Australia-New Zealand during the Korean War," in *The Korean War 1950-1953: A 50 Year Retrospective* (2000), 6; and Forbes, *Korean War*, loc. 346.

11   Robert O'Neill, *Australia in the Korean War 1950-1953* Vol. II, *Combat Operations* (Canberra: AWM, 1985), 9.

12   Andrew Salmon, *Scorched Earth, Black Snow: Britain and Australia in the Korean War* (London: Aurum, 2011), 149; and Forbes, *Korean War*, loc. 362.

13   Richard Trembath, "'But to this day I still ask myself, why did I serve in Korea?': The Formation of K Force," in *The Korean War 1950-1953: A 50 Year Retrospective* (2000), 4.

14   Trembeth, "Formation of K Force," 4; and O'Neill, *Combat Operations*, 7.

15   Trembeth, *A Different Sort of War*, 91; and Forbes, *Korean War*, loc. 379.

16   Franklin B. Cooling, "Allied Interoperability in the Korean War," *Military Review* 63, No. 6 (June 1983): 28; Cyril N. Barclay, *The First Commonwealth Division: The Story of British Commonwealth Land Forces in Korea, 1950-1953* (Aldershot: Gale & Polden, 1954), 12; and Smurthwaite and Washington, *Project Korea*, 14.

17   Anthony Farrar-Hockley, *The British Part in the Korean War* Vol. I, *A Distant Obligation* (London: HMSO, 1990), 145.

18   Peter Lowe, "An Ally and a Recalcitrant General: Great Britain, Douglas MacArthur and the Korean War, 1950-1," *English Historical Review* 105, No. 416 (July 1990): 635.

19   Jeffrey Grey, *The Commonwealth Armies and the Korean War: An Alliance Study* (Manchester: Manchester University Press, 1988), 135; "Discussion on Military Medical Problems in Korea," *Proceedings of Royal Society of Medicine* 46 (June 10, 1953): 1037; O'Neill, *Combat Operations*, 243.

20   O'Neill, *Combat Operations*, 289.

21   Directorate of History and Heritage, Department of National Defence, *Canada and the Korean War* (Canada: Art Global, 2002), 38; Formation of Commonwealth Division for Korea, c 1950, The National Archives at Kew (Hereafter cited as TNA): WO 216/341; Robert J O'Neill, *Australia in the Korean War 1950-1953*, Vol I: *Strategy and Diplomacy* (Canberra: Australian War Memorial and the Australian Government Publication Service, 1981), 93; and Grey, *The Commonwealth Armies*, 100-101.

22   For example, the Australians complained that the Canadians frequently failed to conduct sufficiently aggressive patrols or dominate No Man's Land during the static phase of the war. Blaxland, "The Armies of Australia and Canada..." 29.

23   "Two Years in Action," *Crown News* (July 28, 1953), AWM: AWM85, Item No. 3/5.

24   John C. Blaxland, "The Armies of Canada and Australia: Closer Collaboration?" *Canadian Military Journal* (Autumn 2002): 46; P. Devine, '*La Belle Alliance* — Lessons for Coalition

Warfare from the Korean War 1950-1953,' *ADFJ* No. 118 (May/June 1996): 56.

25 Commanding General's Journal, July 24, 1951, Marshall Library, General Van Fleet Papers, box 81, folder 9; Johnston, *A War of Patrols*, 132; and Devine, "*La Belle Alliance*," 51.

26 For further information on common traditions and similarities between Commonwealth militaries, see Douglas E. Delaney, *The Imperial Army Project: Britain and the Land Forces of the Dominions and India, 1902-1945* (Oxford: Oxford University Press, 2017).

27 Slim, *Defeat into Victory* (London: Cassell, 1956), 541-542.

28 O'Neill, *Combat Operations,* 200; and "The Battle of Maryang San, October 3-8, 1951," AWM, http://korean-war.commemoration.gov.au/china-intervenes-in-korean-war/battle-of-maryang-san-operation-commando.php (Accessed August 22, 2015).

29 "Appendix 10: Operation by 3rd Battalion Royal Australian Regiment (3 RAR)—October 2-8, 1951 'Commando' I and II," 3 RAR Unit War Diaries, October 1951, AWM85: Item 4/34.

30 Francis Hassett, "The Military Team," in *Korea Remembered: The RAN, ARA and RAAF in the Korean War*, eds. Maurie Pears and Fred Kirkland (NSW: Doctrine Wing of Combined Arms Training and Development Centre, 2002).

31 *Battle of Maryang San* map image supplied by the Australian War Memorial, https://www.awm.gov.au/exhibitions/korea/operations/maryang_san/ (accessed January 18, 2016).

32 E.J. Mulholland, "The War in Korea: Summary of Events

May 1951 to July 1953," *Australian Army Journal (AAJ)* No. 53 (October 1953): 5-9.

33 Grey, *An Alliance Study*, 142.

34 Directorate of Infantry, Army Headquarters, "Operation Commando—Korea, October 2-8, 1951," *ADFJ* 115 (November/December 1995): 49-54.

35 Forbes, *Korean War*, loc. 5341.

36 "Interview with Francis Hassett (Battle of Maryang San)," April 20, 1994, AWM: F04833

37 "Appendix 10: Operation by 3 RAR—October 2-8, 1951 'Commando' I and II," 3 RAR Unit War Diaries, October 1951, AWM85: Item 4/34.

38 Philip Oliver Hobson, *Untitled*, July 1951, From AWM: Photograph Collection, ID No. HOBJ2314, https://www.awm.gov.au/collection/HOBJ2314/ (accessed October 30, 2015).

39 O'Neill, *Combat Operations*, 172-173; and Forbes, *Korean War*, loc. 5307.

40 Francis Hassett, "Military Leadership," *ADFJ* 148 (May/June 2001): 56.

41 Interview with Jeffrey James Shelton (As major officer commanding A Company 3 RAR interviewed by David Chinn), AWM: S02291.

42 Alberic Stacpoole, "Obituary: Brigadier George Taylor," *Independent* (September 18, 2011).

43 "Interview with Francis Hassett (Battle of Maryang San)," April

20, 1994, AWM: F04833.

44 Grey, *Commonwealth Armies*, 105.

45 Farrar-Hockley, *An Honourable Discharge*, 366.

46 Grey, *Commonwealth Armies*, 105.

47 "Interview with Francis Hassett (Battle of Maryang San)," April 20, 1994, AWM: F04833.

48 Alberic Stacpoole, "Obituary: Brigadier George Taylor," *Independent* (September 18, 2011).

49 "Interview with Francis Hassett (Battle of Maryang San)," April 20, 1994, AWM: F04833.

50 "Appendix 10: Operation by 3 RAR—October 2-8, 1951 'Commando' I and II," 3 RAR Unit War Diaries, October 1951, AWM85: Item 4/34; and O'Neill, *Combat Operations*, 184.

51 O'Neill, *Combat Operations*, 184.

52 Jeffrey James Shelton, *Untitled*, From AWM: Photo Collection, ID No. P02208.001, https://www.awm.gov.au/collection/ P02208.001/ (Accessed October 30, 2015).

53 Francis Hassett, "The Military Team," in *Korea Remembered: The RAN, ARA and RAAF in the Korean War*.

54 In Commonwealth militaries, majors command companies. While on duty in Korea, several of Hassett's rifle company commanders were granted temporary rank where they did not hold substantive major rank. "Interview with Francis Hassett (Battle of Maryang San)," April 20, 1994, AWM: F04833.

55 Jack Gerke, "Maryang San—Charlie Company," in *Korea Remembered: The RAN, ARA and RAAF in the Korean War*, eds. Maurie Pears and Fred Kirkland (NSW: Doctrine Wing of Combined Arms Training and Development Centre, 2002); Maurie Pears, "Recollections of War," in *Korea Remembered: The RAN, ARA and RAAF in the Korean War*, eds. Maurie Pears and Fred Kirkland (NSW: Doctrine Wing of Combined Arms Training and Development Centre, 2002); Peter Brune, *Those Ragged Bloody Heroes: From the Kokoda Trail to Gona Beach 1942* (UK: Allen & Unwin, 2005); "Interview with Francis Hassett (Battle of Maryang San)," April 20, 1994, AWM: F04833; and "Timeline," Henry William Nicholls, http://hwnicholls.info/timeline/ (Accessed October 21, 2015).

56 "Interview with Francis Hassett (Battle of Maryang San)," April 20, 1994, AWM: F04833.

57 "Jeffrey James Shelton DSO MC (Rtd) as major Officer Commanding A Company 3 RAR interviewed by David Chinn MBE (Rtd)," AWM, https://www.awm.gov.au/collection/S02291/ (Accessed October 22, 2015).

58 William Keys, "Operation Commando," in *Korea Remembered: The RAN, ARA and RAAF in the Korean War*, eds. Maurie Pears and Fred Kirkland (NSW: Doctrine Wing of Combined Arms Training and Development Centre, 2002).

59 "Interview with Francis Hassett (Battle of Maryang San)," April 20, 1994, AWM: F04833.

60 Keys, "Operation Commando," in *Korea Remembered*; Peter Francis Kenny, *We Who Proudly Serve*, s.v. Hardiman, Jack (Australia: Xlibris, 2015).

61 "Interview with Francis Hassett (Battle of Maryang San)," April 20, 1994, AWM: F04833.

62 Gerke, "Maryang San—Charlie Company," in *Korea Remembered*.

63 Hassett, "The Military Team," in *Korea Remembered*.

64 Ibid.

65 Arthur Stanley, "Duty First," in *Korea Remembered: The RAN, ARA and RAAF in the Korean War*, eds. Maurie Pears and Fred Kirkland (NSW: Doctrine Wing of Combined Arms Training and Development Centre, 2002).

66 Jack Gerke as quoted in Bob Breen, *The Battle of Kapyong* (Australia: HQ Training Command, 1992).

67 "Honours and Awards: Henry William Nicholls," AWM, https://www.awm.gov.au/people/rolls/R1517693/ (accessed October 21, 2015); "Honours and Awards: Jack Gerke," AWM, https://www.awm.gov.au/people/rolls/R1547200/ (accessed October 20, 2015); "Honours and Awards: Jeffrey James Shelton," AWM, https://www.awm.gov.au/people/rolls/R1526136/ (accessed October 20, 2015); and "Honours and Awards: Jack Hardiman," AWM, https://www.awm.gov.au/people/rolls/R1531583/ (accessed October 21, 2015).

68 Taylor as quoted in O'Neill, *Combat Operations*, 200.

69 "Stalemate, the War in 1952-1953: Overview of the war from 1952," AWM, http://korean-war.commemoration.gov.au/stalemate-in-korean-war-1952-1953/index.php (Accessed August 22, 2015) (accessed December 19, 2015).

70 1 RAR Unit War Diaries, September 1952, AWM85: Item 2/14.

71 William C. Johnston, *A War of Patrols: Canadian Army Operations in Korea* (Vancouver: UBC Press, 2003), 303.

72 HQ 28 Brigade, Notes on Patrolling in Korea, April 1953, Canadian Directorate of History and Heritage (DHH) 410B25.019 (D238).

73 Lieutenant McCully, 3 RAR Training Instruction No. 5, 3 RAR Unit War Diaries, November 1952, AWM85, 4/47.

74 Daly had also spent time at several different British military schools including the Staff College at Haifa, Staff College at Camberley and the Joint Services Staff College. Brigadier Wilton attended the Imperial Defence College in the late 1940s. Jeffrey Grey, *A Soldier's Soldier: A Biography of Lieutenant General Sir Thomas Daly* (Melbourne: Cambridge University Press, 2013), 81-82; O'Neill, *Combat Operations,* 219, 227, 230, 286; and "Thomas Joseph Daly," AWM, https://www.awm.gov. au/people/P10676275/ (accessed October 20, 2015).

75 Clinton J. Ancker III, "The Evolution of Mission Command in US Army Doctrine 1905 to the Present," *Military Review* (March/April 2013): 42-52; and Eitan Shamir, "The Long and Winding Road: The US Army Managerial Approach to Command and the Adoption of Mission Command (Auftragstaktik)," *Journal of Strategic Studies* 33 (2010): 646.

76 Shamir, "Long and Winding Road," 646.

77 Ibid.

78 Johnston, *A War of Patrols,* 325.

79 Grey, *Commonwealth Armies,* 140.

80 Historically, national commanders operating within a

multinational coalition have been issued with similar directives to protect their command. For example, Canadian officer General Sir Arthur Currie had similar authority during WWI, as did Australian Field Marshal Sir Thomas Blamey during WWII. A.J.H. Cassels, 1 Commonwealth Division Periodic Report, May 1- October 15, 1951, National Archives of Australia (Hereafter cited as NAA): CRS A2107, item K11.09; Doug Delaney, email message to author, September 2015.

81 West, 1 Commonwealth Division Periodic Report, April 1, 1953, (TNA): WO 308/27.

82 Ibid.

83 Devine, *La Belle Alliance*," 51.

84 J.M. Rockingham, Recollections of Korea, August 1975, Library and Archives Canada (LAC), MG 31 G12, 29-30.

85 Johnston, *A War of Patrols*, 148.

86 Author's interview with Field Marshal (retired) A.J.H. Cassels, September 2, 1994 as quoted in David J. Bercuson, *Blood on the Hills: The Canadian Army in the Korean War* (Toronto: University of Toronto Press, 1999), 131-132.

87 Howard G. Coombs, *Canada's Army and the Concept of Maneuver Warfare: The Legacy of the Twentieth Century* (Fort Leavenworth, Kansas: US Army Command and General Staff College, 2002), 23.

88 Author's interview with Field Marshal (retired) A.J.H. Cassels, September 2, 1994 as quoted in David J. Bercuson, *Blood on the Hills: The Canadian Army in the Korean War* (Toronto: University of Toronto Press, 1999), 131-132.

5

A LONG BRIDGE IN TIME: THE 1ST AUSTRALIAN TASK FORCE IN VIETNAM VIA MALAYA AND BORNEO

Dr. Bob Hall

Australian and New Zealand forces forming the 1st Australian Task Force (1ATF) arrived in Phuoc Tuy Province, Republic of Vietnam, in late May 1966. This commitment lasted until October 1971 when the task force was withdrawn. The force initially stood at just over 4,000 men of whom the main combat elements were two Australian infantry battalions, one Special Air Service squadron, an armored personnel carrier (APC) squadron, and a field artillery regiment consisting of three batteries totaling eighteen 105mm guns. 1ATF was later joined by engineers, signalers, army aviation, intelligence, psychological operations, and logistical support units. Number 9 Squadron, Royal Australian Air Force – initially equipped with eight, later sixteen Iroquois helicopters – was based at Vung Tau and provided helicopter support to the task force though they were not part of 1ATF. A New Zealand infantry company joined the Task Force in May 1967, later being joined by another. The task force additionally increased its combat capability with the addition of a third Australian infantry battalion and a squadron of Australian Centurion tanks in early 1968. Referring to the range of capabilities under his command rather than to its manpower strength, one task force commander described his headquarters as "virtually a small divisional headquarters."[1] 1ATF numbered approximately 5,100 men at its peak. It was a small and largely self-contained force.

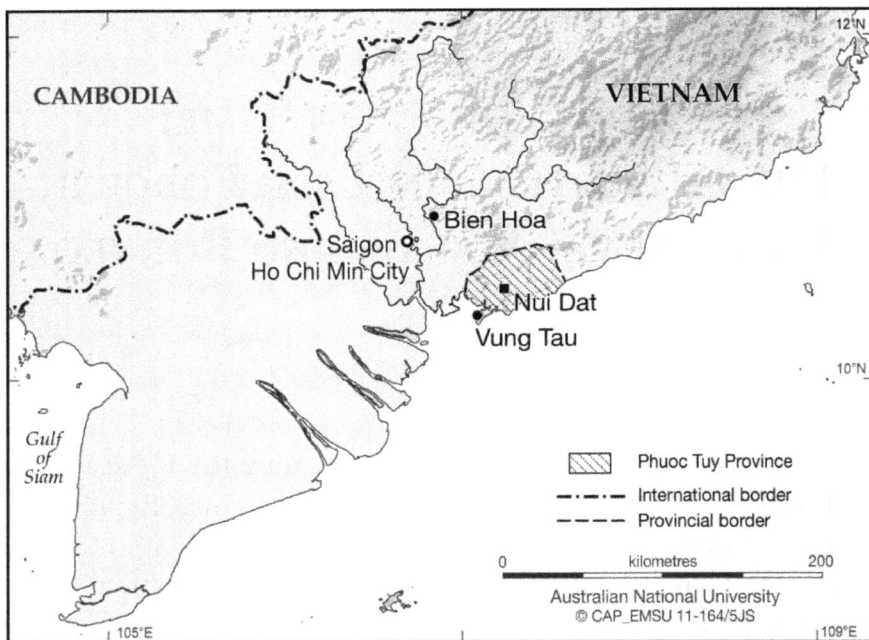

Map 5-1: Southern Republic of Vietnam showing location of Phuoc Tuy Province[2]

MISSION COMMAND IN AUSTRALIAN DOCTRINE

Australian doctrine for the conduct of counterrevolutionary warfare (CRW) (a term regarded at the time as synonymous with counterinsurgency) set out the way in which the Australian task force was expected to conduct operations against the Viet Cong and People's Army of Vietnam operating within its assigned area of operations (AO).[3] The doctrine had its roots in British Commonwealth operations during the Malayan Emergency and Indonesian-Malaysian *Konfrontasi* (also known as the Borneo

Confrontation).[4] Australian forces had participated in both campaigns and although there is debate regarding the extent to which individuals had studied the published CRW doctrine prior to their arrival in Vietnam, commanders at the battalion level had developed a familiarity with counterinsurgency operations that was readily transferred to Vietnam. The doctrine placed emphasis on conducting pacification operations aimed at separating the civilian population from the enemy. Concentration of the Australian force for battle against battalion-sized enemy formations was not foreseen. Doctrine instead envisaged a campaign of company, platoon, and smaller patrols searching for the enemy, conducting ambushes, and slowly whittling the enemy's strength in close collaboration with indigenous forces and local government and military authorities. This type of campaign was thought to raise particular command challenges. CRW doctrine noted:

> high morale is particularly important in counterinsurgency operations where the *normal exercise of command* is difficult and so much depends on the aggressive action of individuals and sub-units.[5]

Counterinsurgency was seen as requiring increased devolution of command decision-making down to sub-unit and patrol commanders.

Australian Army doctrine argued that counterrevolutionary warfare was complex; military aims might sometimes be subordinate to political, social, or psychological requirements. In such circumstances soldiers were to develop a clear understanding of the "military sub-aims or ingredients" of broader political, social, or psychological objectives. Ideally, the political plan established the overarching mission and military operations designed by military commanders were to fit within and complement this broader political plan.[6] It was expected that a coordinating committee (called the Area War Executive Committee) comprising civil and military authorities including the province chief; his advisers on such fields as health, transport, public utilities, and labor; senior police officer; commander of local military forces; representatives of other national groups; and commander of Australian forces and his advisers would be formed. Under the chairmanship of the province chief, this committee would provide direction on complex political and social issues while local military commanders conducted their operations in accordance with committee policies.[7]

The ideas expressed in Australian counterrevolutionary warfare doctrine influenced operations in several ways. First, they acknowledged that command would require devolution to sub-unit and patrol commanders. Second, although the appearance of the term "mission command" remained some years off, they hinted that military commanders should be free to determine how they conducted their operations while conforming to coordinating committee intentions. Third, although conditions precluded counterrevolutionary warfare doctrine ever being fully implemented in Vietnam, these ideas guided the way 1ATF preferred to conduct operations. It was an approach often at odds with several (though not all) US commanders.

A second Australian doctrinal pamphlet had more to say about the exercise of command within the context of a senior military commander's intentions. *The Battalion*, part of the Infantry Training doctrine series, stated, "the battalion commander must not usurp the functions of his subordinates; in modern war greater reliance must be placed on commanders at all levels."[8] *The Battalion* envisaged a narrow range of circumstances in which a form of mission command might be possible, for example, when communications between the battalion commander and his subordinates were cut. It recommended that the battalion commander should issue an operation instruction to his subordinates before each operation to account for such situations.

However, the battalion commander was to exercise personal control when communications allowed.[9]

The Battalion stressed clarity in orders and instructions to ensure that subordinate commanders conformed to their senior commander's intentions:

> An operational instruction must make the future intentions of the battalion commander so clear that the sub-unit commander will instinctively take the correct course of action. It must contain all available information about the enemy and the battalion commander's future plans. It may include alternative plans in order to forestall alternative moves and actions of the enemy.
>
> In general, it must be made absolutely clear to the recipient:
> - What the battalion commander is thinking and what he is going to do.
> - His own task or tasks.[10]

The commander's intentions were to be made known down to the lowest level of command so that the most junior commander would know what was expected of him and what action he must take to fulfil the commander's intentions.[11] *The Battalion* recognized that the achievement of this ideal was dependent on orders passed down the chain of command. Orders were to clearly convey "what the battalion commander intends to do, how he intends to do it, and what part they themselves have to play."[12] Conveying the commander's intent was especially important at the section (squad) or platoon level where there was a high risk of commanders becoming casualties, a point also acknowledged in 1ATF standing operating procedures.[13] Under such circumstances, it was important to ensure that each soldier understood his role in the execution of the commander's plan so that he could carry it out in the event his commander became a casualty.

The doctrine also recognized that departures from orders might become necessary—even advisable—under some circumstances. Unexpected problems or opportunities often emerge during combat. This was especially so in tropical warfare where the terrain, lack of road infrastructure, dense foliage, and changing weather conditions could lead to rapid changes in the threat posed by the enemy, opportunities for exploitation, and interruptions to communications. Operations in tropical Southeast Asia could also create unexpected changes in organization, equipment, mobility, and the extent of force dispersion, demanding flexibility and adaptability. But the doctrine urged that in considering the possible need to diverge from orders, the commander should remember that:

- A formal order is never departed from either in letter or spirit if the issuing officer is present, or there is time to report to him and await a reply without losing an opportunity or endangering the force concerned.
- If this is not possible, a departure in either letter or spirit is only justified if the subordinate assuming the responsibility bases his decision on facts which could not have been known to the officer who issued the order and is satisfied that he is acting as his superior would order him to act were he present.
- If a subordinate neglects to depart from his orders when such departure is clearly demanded, he will be held responsible for any failure that may ensue.
- Should a subordinate find it necessary to depart from an order, he is to inform the originator and the commanders of any neighboring forces likely to be affected as soon as possible.[14]

Therefore, although *The Battalion* seemed to acknowledge the existence of what we now call mission command, the circumstances under which mission command might be applied were heavily prescribed. Rather than encouraging mission command for its own sake, the doctrine seemed to grudgingly concede that it might be necessary from time to time under exceptional circumstances. As will be explained later, this grudging admission of mission command can be seen as a transitional phase, a tentative step towards a readier acceptance of mission command as the norm rather than an exception to be avoided if possible.

MISSION COMMAND AND TRAINING

The Australian Army's system of rotating whole infantry battalions through one-year tours in Vietnam had an impact on mission command to the extent it was practiced within the Australian task force. In the small Australian Army of only nine regular infantry battalions, most officers of captain rank or above and most senior NCO's knew—and in many cases had served with—every other infantry officer or NCO of the same or higher rank. Most had served in the Malayan Emergency, *Konfrontasi*, or both. By 1969 many officers and NCOs were on their second or third tour in Vietnam. My own battalion; the 8ᵗʰ Battalion, the Royal Australian Regiment (8RAR); was typical. Of the nine majors who served in 8RAR, eight had served in Malaya (or Malaysia) and four had previously served in Vietnam. Three had served in other parts of the region including Sabah, Sarawak, Brunei, or Cambodia. Five had served in Papua New Guinea.[15] A similar level of experience existed at the platoon level. My platoon sergeant, two of my three section leaders and some of my soldiers were on their second tour of Vietnam. Even if they did not have previous combat service, many NCOs had known each other and served together over many years. They were very familiar with counterinsurgency tactical techniques favored by the Australian Army.

Battalions warned for Vietnam service were brought up to their full complement of officers, NCOs, and soldiers and began an intensive training program usually beginning about ten months before their deployment. Training progressed through individual, section, platoon, company, and finally battalion level. Companies rotated through an intensive training program at the Jungle Training Centre in Canungra where instructors with recent operational experience from Vietnam trained them in the tactics of jungle and counterinsurgency warfare. Many company commanders held informal tactical exercises without troops or "quick decision exercises" in which junior officers and NCOs discussed how they would deal with tactical problems. These familiarized company officers and NCOs with their colleagues' tactical thinking. Before its departure for Vietnam, each battalion was put through a major exercise under the gaze of experienced exercise control staff.[16]

The officers and NCOs who had steered their battalion through this training process also tended to stay with the battalion throughout its one-year tour in Vietnam. When they arrived in Vietnam, there was therefore already a well understood way of doing things within the unit. Furthermore, battalions, companies, and even platoons had often developed detailed standing operating procedures (SOPs) that codified these ways of doing things although at platoon level these were informal, amounting to a series of understandings about how the details of tactical procedures would be carried out.

Although company commanders therefore usually had a good understanding of the capabilities of their subordinates, this could be further enhanced during operations by attaching company headquarters to one of the three platoons. This gave the company commander an opportunity to observe at close hand how his subordinates within the platoon were coping. Platoons judged least ready for independent patrols might find company headquarters regularly attached to them.[17]

One of the advantages the Australian Army derived from its unit rotation system during the Vietnam War was that it revealed the pattern of performance during the course of a one-year deployment.

In 1971 the Australian Army Operational Research Group studied the distribution of 1ATF accidents. The study revealed a distinct pattern. In the first three months of each infantry battalion's deployment to Vietnam there was a peak of accidents. By the fourth month, units had developed combat experience and had adapted to the combat environment. They had developed "combat savvy." The frequency of accidents fell by more than half. The rate at which accidents occurred remained stable until months eight and nine when there was a moderate "bump" in the accident susceptibility curve before the pattern of accidents settled back to a very low level until the unit withdrew from Vietnam.[18] The bump was attributed to soldiers reaching "an emotional low" in the eighth and ninth months, making them more susceptible to accidents. A similar pattern was seen in the incidence of friendly fire clashes and accidental weapons discharges.[19] Such patterns may have future mission command implications. Careful monitoring of the accident susceptibility curve could signal when subordinates require greater clarity of orders and closer supervision.

Intensive pre-deployment training built a junior commander's confidence in his performance and that of colleagues. Confidence tended to develop further once junior commanders experienced their first taste of combat. 1ATF dominated the enemy in Phuoc Tuy Province. Patrols saw the enemy first and initiated contact in the overwhelming majority of all contact types.[20] Loss ratios heavily favored 1ATF in all except bunker system attacks.[21] Ready availability of artillery and air support further boosted the confidence of infantry patrol commanders. As enemy capability in the 1ATF AO waned, the Task Force was able to deploy platoon strength patrols. Half-platoon and sometimes section-strength patrols became more commonplace from mid-1969. One sign of increased confidence in junior commanders' tactical ability: platoon strength patrols often operated up to three kilometers from parent company headquarters for periods of five or six days, re-joining company headquarters only for resupply. Dispersed operations made "mission command" essential, doctrine notwithstanding.

I usually commanded any platoon or half-platoon patrol operating away from company headquarters. After receiving broad direction from my company commander, usually to either search for the enemy or conduct ambushes in a particular area, I was often left to conduct my patrol in my own way. I planned my own search pattern, decided which track should be ambushed, set ambushes to my preference, decided how and by whom it would be sprung, and decided whether the ambush would remain in place or move to a new position after execution. I reported my patrol location hourly to company headquarters. My company commander had a brief radio conversation with each of his patrol commanders each day around nightfall, reviewing the day's activities and discussing future plans. I was expected to say what I thought my patrol should do based on my assessment of the situation. I might argue that there was a high possibility of a contact and that my patrol should stay in ambush if I had found a well-used track showing fresh footprints. On the other hand, I might suggest that the patrol should search a new area. Our company commander provided coordinating advice. He might direct that I stay within a specified distance of neighboring patrols so that they (or we) could be quickly reinforced if necessary.

APPROACHES TO THE WAR

Both the US and Australian ground combat forces arrived in Vietnam at a time when the South Vietnamese government faced imminent collapse as a result of enemy battlefield victories. Allied intervention came at a time when the enemy mounted attacks of multi-regiment if not division size.

The 1st Australian Task Force was under command of Commander, Australian Force Vietnam (COMAFV) who, in turn, was subordinate to General William C. Westmoreland, Commander, United States Military Assistance Command, Vietnam (COMUSMACV). However, 1ATF was under operational control of the US Army's Commanding General (CG), II Field Force Vietnam (II FFV) who was responsible for operations throughout III Corps Tactical Zone (III CTZ). Commander, 1ATF was obliged to accept orders and directives from CG, II FFV and conduct operations in accordance with his intentions.

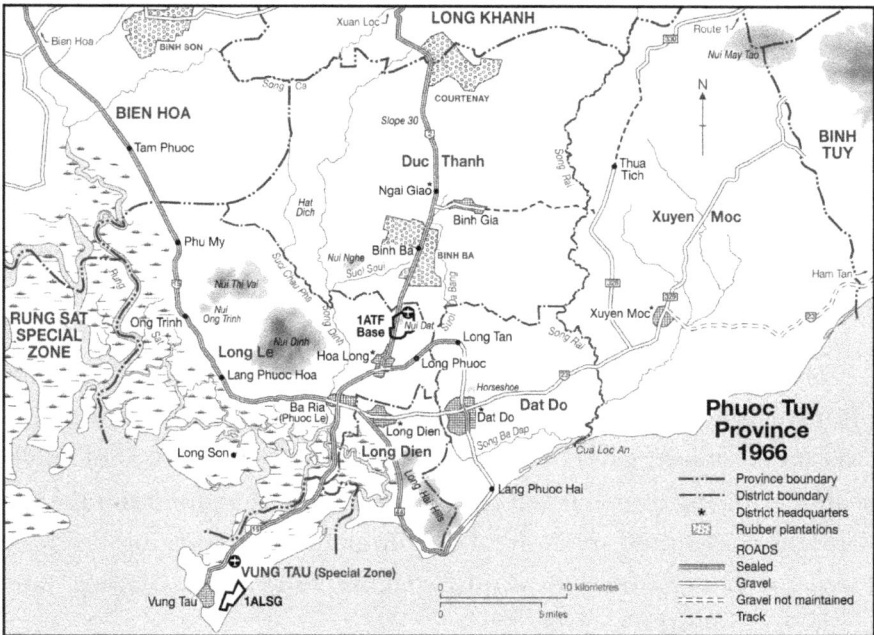

Map 5-2: Phuoc Tuy Province, Republic of Vietnam[22]

Australian battalion commanders felt that they had a good grasp of counterinsurgency warfare. Australian doctrine made security of the population top priority. This approach was applied successfully in Malaya and North Borneo. But US senior commanders constrained Australian commanders from applying preferred techniques as fully as they would like. The campaign in Phuoc Tuy Province was more like a classic counterinsurgency, consisting of numerous small scale contacts, than the war of big battles fought by US forces in I CTZ and the western provinces of III CTZ. While it is not surprising that the war of big battles therefore tended to dominate USMACV thinking, this approach generally failed when applied in Phuoc Tuy Province where the enemy sought to avoid major engagements.

Australian commanders felt themselves obligated to comply with orders emanating from CG II FFV whether or not they agreed with the USMACV approach. As COMAFV Major General Robert Hay explained to the Australian ambassador in Saigon in 1969,

> You will appreciate that the Australian military effort is under the operational control of MACV, and AFV quite rightly is obliged to follow US priorities. It would be unthinkable for Australia to contemplate developing independent operations on national lines.[23]

However, within their superior commanders' intent, and often in the absence of large enemy formations within their AO, Australian commanders often applied tactical techniques of small scale patrolling more suited to counterinsurgency operations. This sometimes resulted in senior US commanders criticizing the task force for being overly cautious, seeking to avoid casualties, and achieving a poor loss ratio.[24]

US strategy underwent a shift in direction following General Abrams' July 1968 replacement of General Westmoreland. The Tet Offensive of 1968 had reduced enemy capability, enabling Abrams to replace Westmoreland's attrition approach with a pacification strategy. Security of the population became the top priority. Upgrading ARVN capabilities was second priority; bringing enemy main force elements to battle slipped to third.[25] The change allowed 1ATF to apply its counterinsurgency doctrine more freely.

The change of strategic policy was not fully embraced by some senior US commanders. Lieutenant General Julian Ewell; CG, II FFV from April 1969 to April 1970; was one resisting change. Ewell's first directive to subordinate commanders outlined Abrams' new priorities but then dictated a return to the attrition strategy under Abrams' predecessor. Objectives sought to maximize enemy casualties rather than secure the population:

> Our immediate goal is to attrit the enemy at a rate of 6,000 per month by the end of April. This figure is the break even point and is merely the entry price. Within the units, we must put the requirement on them to produce results.[26]

Ewell's directives show a commander intent on directing his subordinates in detail, instructing them to increase enemy casualties via more "company days in the field" with "30 to 40% of company effort" on night offensive operations and ambushes. Directives further dictated policies regarding zeroing of rifles, marksmanship training, ambush techniques and patrolling, and how best to integrate new reinforcements.[27] A later memorandum urged subordinate commanders not to employ their troops on population security tasks "unless it's quite clear that the hamlet will be lost unless we step in with US or ARVN forces;" if committed, they were to obtain an "early and firm release date."[28]

Senior Australian commanders were scornful of Ewell's approach. Major General (MAJGEN) Stuart Graham, former 1ATF commander and Deputy Chief of the (Australian) General Staff, wrote to MAJGEN Hay, COMAFV in May 1969 saying that Ewell's April memorandum "depressed" him:

> One could take issue with it on many basic points and indeed it could have been written by a French general 20 years ago. What really intrigued me, though, was that it started off giving

top priority to pacification and finished up with the old (and to my mind, discredited) message "get out and kill them bums."[29]

Junior Australian commanders also had to deal with Ewell. Ewell visited Lieutenant Colonel Ron Grey at his 7RAR headquarters in the field during April 1970. Ewell was very critical of Australian operations and continued to emphasize the importance of statistics and body count. In his history of 7RAR in Vietnam, Michael O'Brien wrote, "the atmosphere [of the meeting] was cold and one of overbearing disagreement."[30]

One operation in particular irked Australian commanders. Conducted from January 15-26, 1970, Operation Matilda was initiated by HQ 1ATF in response to pressure from Ewell to mount a large-scale operation aimed at bringing enemy main force elements to battle. It involved forty-four armored vehicles including Centurion tanks and APCs, artillery, a rifle company, and helicopter and fixed wing air support on a 240 kilometer drive through parts of neighboring Binh Tuy and Long Khanh Provinces. The operation netted just two Viet Cong killed in action despite the massive resources committed.[31] Operation Napier, an infantry heavy operation covering a similar period with less lavish support and employing the infantry patrolling and ambushing techniques preferred by Australian commanders, killed thirty-nine enemy, wounded thirteen, captured another thirteen, caused one to rally, and detained sixty-nine.[32]

Ewell's directive command style rankled many Australian commanders. Fortunately, Ewell was atypical. The command style of Lieutenant General Frederick C. Weyand, Ewell's predecessor as Commanding General II FFV, stood in sharp contrast to Ewell's. For example, on the eve of the 1968 Tet Offensive, Weyand gave the Australian Task Force commander wide discretion when deploying task force elements north east of Saigon; 'Waving vaguely at the map [Weyand] casually asked the commander if he would mind bringing his task force up to where his hand lay'.[33] The task force commander was taken aback by Weyand's informality, the brevity of his orders and the flexibility of action they implied. Lieutenant General Michael Davison who replaced Ewell in April 1970, also stood in contrast to Ewell. Australian commanders found Davison far more congenial, a man who allowed subordinate commanders greater initiative within his intent. However, Ewell's highly directive style of command provides an insight into an important characteristic of mission command; the marked difference between Ewell, Weyand and Davison indicates that effective mission command may depend as much on the personality of the senior commander as on doctrine and adaptation to operational conditions.

MISSION COMMAND AT THE BATTALION LEVEL

Despite lack of a doctrine encouraging mission command, 1ATF infantry battalion and company commanders tended to assign subordinates a mission without specifying in detail how the mission was to be executed. Writing in the Australian *Army Journal*, Major Reg Sutton, a company commander in 5RAR, noted:

> The company commander in Vietnam has a great deal of flexibility in the command and control of his three platoons. When he is allotted a task in an AO, he can normally operate his three platoons without undue interference from his battalion commander.... The company commander is given a task to perform within the framework of the CO's overall plan.[34]

As mentioned, one of the reasons for this was that 1ATF operations were generally highly dispersed. Platoon, half-platoon, and even section rather than company-strength patrols became increasingly common. The security situation in Phuoc Tuy Province was sufficiently benign by mid-1969 to allow use of half-platoon patrols consisting of twelve or thirteen men equipped with two M60 machineguns and an ANPRC-25 radio set. But this does not imply that Phuoc Tuy was peaceful. Total contacts peaked in 1969 and though some contacts were with large enemy groups—such as the ambush of a 60-man enemy group by a 6RAR platoon-strength patrol on 20 June 1969[35] - most contacts after June that year were with small groups of enemy. In 1969 there were 968 contacts by 1ATF elements numbering less than 17 men—that is, about the size of a half-platoon patrol - compared to 237 contacts by elements greater than half-platoon size. In 1970 the number of contacts involving 17 men or less had shrunk to 568, but those involving more than 16 men had shrunk further to just 115. In this period, most contacts were with enemy groups of 6 men or less, and could usually be dealt with by a half-platoon. Contacts with groups of enemy greater than 35 men averaged about 1 per month through 1969 and 1970. In these circumstances half-platoon patrols became possible and though the potential for such patrols to encounter larger enemy forces than themselves added an element of risk this was deemed acceptable by 1ATF commanders.[36] By May 1970 the Cambodian incursion led to a further improvement in the security situation in Phuoc Tuy province where local Viet Cong units were at the end of an increasingly tenuous supply line. Though contacts with the enemy continued, half-platoon patrols became the norm. Platoon commanders or platoon sergeants commanded the patrols. Company headquarters often provided an additional ambush patrol, bringing the total of separate combat elements within a rifle company to seven. Together with other battalion and supporting elements, some battalion command posts were at times controlling up to forty-eight individual tactical elements.[37] These could be widely dispersed over many kilometers. For example, 8RAR deployed some 140 separate ambush patrols across an area of nearly eighty square kilometers that included jungle, rubber plantation, paddy fields, and villages during Operation Phoi Hop from April 7-19, 1970.[38] Such broad distribution precluded detailed centralized command. Patrol commanders conducted operations within their superior commanders' intent while being responsive to local tactical circumstances. Responsibility for planning operations, developing deception plans, and administration and logistical arrangements usually devolved to battalion and company commanders, albeit with coordination provided by task force and battalion headquarters. Battalion commanders were given a relatively free rein. Lieutenant Colonel Keith O'Neill, CO of 8RAR, recalled:

> There was no task force concept [of operations], so battalions were let go to do what they wanted up to a point. As long as they were chasing the enemy and showing a certain amount of aggression they'd be let go. No-one pulled you up. But people would criticise you if you failed or something went wrong.[39]

In April 1970, O'Neill used this freedom of action to conduct an "experimental" operation in which dozens of half-platoon ambush patrols were placed inside and around the perimeter of Binh Ba, Suoi Nghe, and Hoa Long villages,[40] a classic pacification tactic unlikely to have the support of Ewell. The operation was a success. There were several contacts with enemy patrols attempting to enter the villages, confusing and disrupting their resupply efforts. 1ATF adapted, shifting focus away from long, arduous patrols attempting to find main force elements in the jungle towards ambushing around the villages, a tactic that had the benefit of forcing the enemy to fight for entry.

Most 1ATF contacts occurred with little more than a few seconds warning. Immediate tasks for the commander were to establish domination of the enemy and then begin to think about what might happen next. Most contacts lasted only a few minutes. Approximately sixty per cent of all 1ATF ambushes, patrol encounters, and security (or defensive) contacts ended in less than ten minutes.[41] Only in bunker system attacks where the enemy had selected and prepared the ground—and therefore felt the battle was likely to flow in his favor—did duration of contacts tend to be significantly longer. Contact lasted over thirty minutes in fifty-eight per cent of bunker system attacks.

Smaller patrols often lacked time to do anything but implement contact drills during their brief encounters. Standardized drills had been rehearsed in training until they became second nature. With the flexibility enjoyed through mission command, many commanders adapted these drills to better suit the tactical situations they encountered in combat. For example, Australian minor tactics doctrine insisted that the response to enemy ambush should be to immediately assault the ambush position while returning a high volume of fire. This was impractical in Vietnam. The advent of the claymore mine meant that a single enemy soldier could deliver massive firepower into the killing ground. It was unwise to be standing when he did so. Many junior commanders therefore modified contact drills to get patrols onto the ground, dispersed and presenting a broad front to the enemy (whether seen or unseen) while returning fire.[42]

Such immediate action drills momentarily took control out of a patrol commander's hands. His next opportunity to influence an engagement often began upon completion of the contact drill. But the opportunity for decisive action was often gone before he could initiate action. In a review of infantry lessons, a conference of former infantry Vietnam War battalion commanders seemed oblivious to this fact. They argued that training over emphasized contact drills and that patrol commanders allowed momentum to stall upon their completion.[43] While Australian doctrine grudgingly acknowledged mission command might be acceptable under limited circumstances, battalion commanders developed the expectation that subordinates should *always* exercise aggressive initiative as specified by their superior commanders' intent and were critical of subordinates failing to do so.

Sometimes conditions did favor junior commanders executing their own plan to achieve their commander's intent. One such case occurred in paddy fields outside the village of Hoa Long in southern Phuoc Tuy Province on the night of August 11-12, 1970. Twenty-four men of 8 Platoon, C Company, 8RAR under the command of Sergeant Chad Sherrin set up an ambush on the approaches to the village, a regular enemy source of supplies. The ambush was part of Operation Cung Chung for which the commander's intent was to interdict enemy entering villages to resupply.[44] Sherrin's patrol saw a group of fifty to sixty enemy approaching in the moonlight about 100 meters distant, tactically spaced, and moving cautiously. They were too far away to engage effectively. Opening fire might have caused the enemy to abort the mission, thus achieving the commander's intent. However, Sherrin and his men held their fire, watching as the enemy filed past and entered Hoa Long. Sherrin re-positioned the ambush to cover the ingress route taken by the enemy. It was risky business in the dark, requiring retrieval and re-setting of claymore mines and re-siting both machine guns and individual riflemen.[45] Meanwhile, back at Nui Dat, the task force commander, commanding officer of 8RAR, and Sherrin's company commander entered into a heated debate about the patrol leader's failure to deny the foe entry into the village.

Carrying heavy packs filled with their resupply and confident that by retracing their route they would have a safe passage back to the jungle, the enemy patrol made their way out of Hoa Long several hours later. They were bunched and vulnerable. The enemy patrol took a route that led them straight towards the waiting men, though not through the killing ground Sherrin had expected they would enter. Sherrin waited until the leading enemy elements had approached to within about four meters of the ambush, then opened fire with banks of claymores followed by machineguns,

M72 rocket launchers, and other small arms. Nineteen enemy were killed, six captured. One later rallied to the government. The patrol also captured the laden backpacks. Sherrin was awarded the Military Medal for the action. It is likely he would have received a very different reward had the enemy patrol taken a different route and his ambush not been successful. While Sherrin's ambush failed to prevent the enemy from accessing the village, it did deny the enemy resupply, meeting the actual intent within what was perhaps a less than perfectly worded senior commander's intent. The enemy resupply party was later found to be from the Chau Duc District Company. Together with earlier 1ATF ambushes, Sherrin's ambush had virtually destroyed the enemy company as an effective fighting force.[46]

Other cases of exercising initiative by junior commanders had less positive outcomes. A sixteen-man half-platoon patrol of 8 Platoon, C Company, 3RAR, was patrolling in search of the enemy on March 20, 1971. The patrol was part of Operation Briar Patch, a search and ambush operation seeking to interdict enemy penetrations into Phuoc Tuy Province. Sounds of chopping, digging, and a dog barking ahead should have alerted the patrol commander to the possibility that an enemy bunker system was nearby. He ordered his troops to drop their field packs and deployed the patrol for an assault. Without conducting a reconnaissance, he opened fire on the still unseen source of the noises with M79 and rifle grenade fire, forfeiting surprise before advancing in assault formation. The enemy responded with command-detonated CHICOM "claymore" fire and small arms, killing the patrol commander, wounding a machine gunner, and pinning the patrol down. Two helicopter gunships called on to provide support could not identify the patrol's position and were unable to provide support; most of the half-platoon's smoke grenades had been left at the start line with the field packs. The few available were insufficient to continuously mark the Australians' position. Enemy fire wounded one of the gunship pilots who later died of his wounds. Supporting artillery fires could not be adjusted because of uncertainty regarding the patrol's location. Lieutenant Colonel Peter Scott, CO 3RAR was airborne above the platoon. Unable to drop additional smoke grenades to the patrol, he provided what advice he could while coordinating reinforcements. Other platoons rushed to the contact in APCs. Enemy pressure eased and the patrol was able to disengage after three hours of fighting. The enemy, estimated to number about sixteen men, had withdrawn in their own time and without any apparent casualties.[47] It is likely the engagement would have gone differently had the patrol commander been under closer supervision.

Scott's efforts to orbit the contact in his helicopter and provide support and guidance for the junior commander on the ground was a response typical of many battalion commanding officers. It reflected their understandable desire to maintain control and help a sub-unit in trouble. But it had negative consequences. It tended to give the senior officer an illusion of better situational understanding than was actually the case. Applying the commanding officer's guidance based on his perspective could be problematic. Scott later acknowledged that he had become too involved in the platoon commander's fight:

> While the desire of airborne elements to assist ground troops in contact by dropping bags of smoke grenades or ammunition is supportable in some circumstances, there is often a down side in interference with the conduct of the immediate battle.[48]

Scott's constant radio communications with the platoon on the ground had prevented the company commander from communicating with his other platoons to organize support for the platoon in contact. Scott had prevented his company commander from making what might have been a decisive contribution to the battle. Unlike as specified by doctrine at the time, senior commanders have to refrain from taking over simply because they can if mission command is to be successfully implemented.

Scott later imposed several guidelines aimed at preventing a recurrence. Half-platoon patrols were not to operate more than 250 meters from a firm base. Platoons were not to be separated by more than 500 meters from each other to facilitate quick reinforcement. Firing at movement or noise without first identifying its source was not to occur. Smoke grenades were always to be carried on the soldier's basic webbing, not in his field pack.[49] For Scott, these instructions set guidelines within which he could continue to exercise mission command.

This incident was one in which prior training failed to deliver in combat. It is noteworthy that this contact occurred within the first month of 3RAR's second tour in Vietnam. The platoon had had only one previous contact with the enemy, a very fleeting and inconclusive contact with a single Viet Cong about three weeks earlier. The platoon was therefore relatively un-blooded. Sherrin's ambush, on the other hand, had occurred at the end of 8RAR's tour. Sherrin's capabilities as a combat leader were well known, especially to his company and platoon commander. His platoon had fought thirteen earlier firefights. As their calm appreciation of the situation and subsequent actions testify, Sherrin and his men were experienced and justifiably confident in their capabilities. Effective mission command requires that senior commanders carefully assess a junior commander's expertise before allowing the individual the freedom to act on his own initiative. This assessment can be based on their junior commander's performance in pre-deployment training and exercises and, once deployed to the combat zone, on that under fire.

The exercise of initiative by junior commanders in Vietnam often had more to do with gaining contact with the enemy than conducting the firefight itself. Commanders at the company and platoon level often had great flexibility regarding tactical decisions like whether and how to receive ration resupply, whether to allow cooking in night bivouacs, or how to conduct a search of a creek line. In an effort to remain undetected, one company commander used an ox cart driven by a Vietnamese soldier to deliver the company resupply in lieu of relying on helicopters.

As we have noted, such mission command doctrine as existed in Vietnam tended to view the approach as exceptional, something to be avoided if possible. Several factors influenced army's views of mission command as 1ATF experience ripened. Australian senior commanders found that they had to conform to what they believed was Westmoreland's inappropriate strategy for the situation in their AO. While 1ATF was required to seek out and destroy enemy main force units, it went about the process in its own way, often coming under criticism from senior US commanders. Task force commanders by and large shielded their subordinates from undue interference from higher. Mission command was readily applied at the battalion level. Highly dispersed operations using platoon and half-platoon patrols left little alternative to allowing proven junior commanders to use their initiative within the bounds of their commanders' intentions.

ENDNOTES

1   D.M. Horner, Australian Higher Command in the Vietnam War, (The Strategic and Defence Studies Centre, Research School of Pacific Studies, Australian National University, 1986), 33.

2   Map courtesy of the Australian National University College of Asia and the Pacific cartographic unit, map number 11-164/5JS.

3   The Division in Battle, Pamphlet no. 11, Counter Revolutionary Warfare, 1965, Military Board, Army Headquarters, Canberra, 1966.

4   Konfrontasi was an attempt by Indonesia to disrupt the federation of Malaya and the north Borneo states of Sarawak and Sabah to form Malaysia. It was the brainchild of President Sukarno of Indonesia. It took place from 1963 to 1966 and initially involved Indonesian "volunteers," many of whom were Indonesian soldiers not in uniform, but later formed units of the Indonesian military. They were opposed by British Commonwealth forces including those from Britain, Malaysia, Australia, and New Zealand. Most of the combat occurred in north Borneo where the Malaysian states of Sabah and Sarawak shared a land border with Indonesian Kalimantan. Commonwealth operations tended to take the form of a series of isolated company strength bases from which infantry patrols sought to find and bring to battle those Indonesian patrols that had crossed into Malaysia. However, the British Director of Borneo Operations mounted secret patrols into Indonesian Kalimantan (known as CLARET operations) to disrupt the Indonesian campaign. See Peter Dennis and Jeffrey Grey, Emergency and Confrontation: Australian Military Operations in Malaya and Borneo 1950-1966, (St Leonards, Allen & Unwin, 1996); and Nick van der Bijl, Confrontation: The War with Indonesia 1962-1966, (Barnsley, Pen & Sword, 2009).

5   Counter Revolutionary Warfare, 33 (emphasis added).

6   Ibid., 28.

7   Ibid., 41. The requirement for these coordinating committees was derived from the experience of the Malayan Emergency. The concept recognizes the primacy of political and social coordination in the conduct of effective counterinsurgency. For a variety of reasons including the ineffectiveness of province administration and enemy penetration of the civil administration, these committees were never able to be implemented by 1ATF in Vietnam.

8   Infantry Training, Volume 4, part 1, The Battalion, 1967, Army Headquarters, Canberra, 1967, 117.

9   The Battalion, 117.

10   Ibid.

11   Ibid.

12   Ibid., 128.

13   1st Australian Task Force, Vietnam, Standing Operating Procedures for Operations in Vietnam, 1-1.

14   The Battalion, 129 and 279.

15   Robert A. Hall, Combat Battalion: The Eighth Battalion in Vietnam (St Leonards, Allen & Unwin, 2000), 9.

16   Ian Kuring, Red Coats to Cams: A History of Australian Infantry, 1788-2001 (Army History Unit, Department of Defence, 2004), 354.

17 The whole of company headquarters was attached with the company commander running his headquarters (HQ) from within the platoon perimeter. Usually the platoon headquarters was a few meters away from company headquarters (CHQ) and the company commander had a lot of opportunity to visit the platoon, talk to the platoon commander and see how he ran things. The company support section (which normally stayed with CHQ to provide security) was usually given a position on the perimeter when CHQ was with a platoon.

18 Australian Army Operational Research Group, Report 1/71, Accidental Casualty Study – South Vietnam, February 1971, 13.

19 Bob Hall, "Accidental Discharges: The Soldier's Industrial Accident in Vietnam and East Timor," Australian Defence Force Journal, no. 149, July/August 2001, 29. Unpublished research by Bob Hall and Andrew Ross on friendly fire incidents shows that these also tend to conform to the described pattern.

20 Bob Hall and Andrew Ross, "Kinetics in counterinsurgency: Some influences on soldier combat performance in the 1st Australian Task Force in the Vietnam War," Small Wars & Insurgencies 21 (September 2010): 510. 1ATF was first to fire in 78% of patrol encounters, 96% of ambushes of the enemy, 82% of bunker system attacks, and 64% of security contacts.

21 Andrew Ross, Robert Hall, and Amy Griffin, The Search for Tactical Success in Vietnam: An Analysis of Australian Task Force Combat Operations (Melbourne, Cambridge University Press, 2015), 94, 134, 163, and 176. Loss ratio was strongly in favor of 1ATF in all but a relatively small number of isolated cases.

22 Map courtesy the Australian National University CartoGIS, "PhuocTuyProvince.tif."

23 Australian War Memorial (hereinafter AWM), series 98, 69/M/7, HQ AFV DO correspondence. Letter, Major General R.A. Hay, COMAFV, to HE R.L. Harry, Australian Ambassador, Saigon, October 4, 1969.

24 Ashley Ekins with Ian McNeill, Fighting to the Finish: The Australian Army and the Vietnam War, 1968-1975, (St Leonards, Allen & Unwin, 2012), 41-43.

25 AWM98, 69/M/7. HQ AFV DO correspondence. Directive issued on April 16, 1969 to 1ATF by CG II FFV.

26 AWM98, R569-1-196, Headquarters, Australian Forces Vietnam, Operation - General II FFV Operational Directives. Memorandum, HQ II FFV, April 16, 1969.

27 Ibid.

28 AWM98, R569-1-196, Headquarters, Australian Forces Vietnam, Operation - General II FFV Operational Directives. Memorandum, HQ II FFV, May 6, 1969.

29 AWM98, item 69/M/7. HQ AFV DO correspondence. Letter, Major General S.C. Graham, DCGS, to Major General R.A. Hay, COMAFV, May 28, 1969.

30 Michael O'Brien, Conscripts and Regulars: With the Seventh Battalion in Vietnam (St Leonards, Allen & Unwin, 1995), 166-7.

31 Ekins with McNeill, Fighting to the Finish, 368-370 and 745. One of the two enemy casualties was killed by Naval Gunfire Support.

32 Ekins with McNeill, Fighting to the Finish, 745.

33 Ian McNeill and Ashley Ekins, On the Offensive: The
    Australian Army in the Vietnam War, January 1967-June 1968,
    Allen & Unwin, Crows Nest, 2003, p. 290.

34 Major R.F. Sutton, "Notes on company operations in Vietnam,"
    Army Journal, no. 262 (March 1971): 14-15.

35 AWM95, item 7/6/22, 10 platoon, D Company 6RAR contact
    report, p. 103.

36 The quantitative data is derived from the 1ATF Combat
    Database, created by Andrew Ross. This database contains 4,665
    combat incidents with up to thirty bits of information relating
    to each incident. The bits of information include the date, time,
    place, friendly unit involved, enemy and friendly strength,
    casualties, duration of contact, and many other factors.

37 Colonel F. Peter Scott, Command in Vietnam: Reflections of a
    Commanding Officer (McCrae, Slouch Hat Publications, 2007), 68.

38 Robert A. Hall, Combat Battalion: The Eighth Battalion in
    Vietnam (St Leonards, Allen & Unwin, 2000), 63.

39 Ibid., 7-8.

40 Major Adrian Clunies-Ross, (ed.), The Grey Eight in Vietnam:
    The History of Eighth Battalion The Royal Australian
    Regiment, November 1969 – November 1970 (Bowen Hills,
    Eighth Battalion RAR, n.d.), 69-72.

41 1ATF Combat Database, created by Andrew Ross.

42 See AWM115, 56, AHQ, Lessons learned by 3RAR in the 1971
    Vietnam Tour [Report] by F.P. Scott, lieutenant colonel, CO

3RAR dated December 29, 1971. In his report, Scott sought input from all the officers and non-commissioned officers of his battalion. His report contains arguments attributed to his subordinates for and against modification of various drills and practices indicating that, within limits, junior commanders in 3RAR had the ability to apply modifications as they saw fit. Another modification to minor tactics was the abandonment of fire and movement at the section (or squad) level. See Brigadier S.C. Graham, "Observations on Operations in Vietnam," Army Journal, no. 235 (December 1968): 5-32.

43 Australian Army, Training Information Bulletin, no. 69, Infantry Battalion Lessons from Vietnam 1965-1971, Headquarters Training Command, 1988, 3-18. This review of lessons made no mention of mission command or the exercise of initiative by junior officers to achieve the commander's intentions.

44 Ekins with McNeill, Fighting to the finish, 456-459.

45 Ibid.

46 Ibid. See also Clunies-Ross, The Grey Eight in Vietnam, 109-117; and Major A. Clunies-Ross, "Ambush by night," Australian Infantry (January 1971): 6-9.

47 AWM279, 723/R5/117, 3RAR Combat Operations After Action Report, Operation Briar Patch.

48 Colin Brewer, Company Commander, 2RAR/NZ, quoted in Scott, Command in Vietnam, 69.

49 Scott, Command in Vietnam, 65-66.

6

THE APPLICATION OF MISSION COMMAND BY THE 1ST BATTALION, THE ROYAL AUSTRALIAN REGIMENT (1 RAR) GROUP ON OPERATION SOLACE IN SOMALIA IN 1993

Lieutenant General John Caligari, AO, DSC, Australian Army (Retired)

INTRODUCTION

In this chapter I describe the application of mission command during the 1st Battalion, The Royal Australian Regiment (1 RAR) Group's preparation, deployment to, and conduct of operations in Somalia from January to May 1993. I was the operations officer (OPSO/S3) of 1 RAR and the 1 RAR Group in Somalia during this period.

While mission command was practiced extensively in 1 RAR and extended well below the level of the CO, his headquarters and his company commanders, the length of this chapter precludes its examination beyond company commander level. I narrowed my focus by restricting my attention to the application of mission command by the battalion commander Lieutenant Colonel David Hurley and the interactions between him [and his headquarters (HQ)] and those four rifle company commanders. I was fortunate in working closely with the CO as his operations officer and therefore being able to watch these interactions firsthand.¹

The Australian Defence Force's (ADF) approach to operations has for a long time been underpinned by maneuver warfare theory and supported by the philosophy of mission command. However, the command philosophy at the time of this deployment to Somalia was described as "directive control,"¹ a somewhat less sophisticated concept, but one that was still very much the basis for the later internationally accepted term of "mission command."² I use the term mission command during the rest of this chapter to remain consistent with the rest of this book.

AUSTRALIA'S SOMALIA DEPLOYMENT

In January 1991, a civil war led to the overthrow of President Siad Barre and the collapse of the Somali state. The country subsequently fragmented into a number of warring factions. To make matters worse, by 1992 the Somali people faced starvation as a result of a cycle of drought and famine. International concern regarding the impact of famine and lack of security led to a limited intervention in April 1992 called United Nations Operation in Somalia (UNOSOM). The failure of UNOSOM to establish an effective ceasefire between warring factions—one allowing distribution of aid—led in turn to Operation Restore Hope, a United States (US)-led, UN-sanctioned peace enforcement operation authorized under Chapter VII of the UN Charter.³ Restore Hope began with the landing of US marines on the Somali coast in December 1992. The Australian contribution was named Operation Solace. 1 RAR Group was assigned the operation, its main body deploying to Somalia in mid-January 1993. It was the first battalion-sized Australian Defence Force deployment outside Australia since the Vietnam War.

---

1   All four of the rifle company commanders assisted in the development of this chapter. The analysis and any shortfalls are my own, however. Nor do I pretend to represent the views of Lieutenant Colonel Hurley but rather only offer my observations regarding our deployment to Somalia.

Early intelligence analysis reflected that the operation was likely to be a low threat undertaking in keeping with what had become the ADF's training and exercise paradigm since the release of the Australian government's White Paper "The Defence of Australia 1987."While instruction at the army's schools had long concentrated on conventional warfare as being the best platform from which to adapt to what the US was at the time calling operations other than war (OOTW), the emphasis on low-level operations and the defense of Australia meant less weight had been placed on armored and other heavy forces. A greater portion of the service was to move to and learn to operate in Australia's expansive, harsh, and sparsely populated north. Greater emphasis was also to be given to the army reserves. Mobility was to be improved by transfer of the helicopters from the Royal Australian Air Force to the army and purchase of a fleet of new soft skinned vehicles was in the offing.[4] There were many similarities between what the 1 RAR Group experienced in Somalia and the low level operations envisaged in the 1987 Defence White Paper and practiced in exercises since its release. 1 RAR would find that Operation Solace included elements of counterinsurgency (COIN), noncombatant evacuation operations (NEO) (referred to by the ADF at the time as services protected evacuations), and military operations on urbanized terrain (MOUT) as well as aid to the civil power (ACP); the last of these particularly in Solace's later stages. 1 RAR participated in two major exercises in 1992 that particularly enhanced our preparation, especially for low-level operations and NEO, which I will elaborate shortly.

CIRCUMSTANCE OF DEPLOYMENT

Before undertaking my analysis of mission command during Operation Solace, it is essential to cover some of the undertaking's background aspects, particularly those regarding the training and exercises conducted by 1 RAR under Lieutenant Colonel Hurley in the twelve months since he took command, the key personalities involved, and circumstances in which we found ourselves on arriving in Somalia.

1 RAR is an infantry battalion in the 3ʳᵈ Brigade located in Townsville, Australia. In 1992, the battalion had four rifle companies, a support company comprised of specialist weapons and combat support platoons, and an administration company. Since the late 1970s the Australian Army had established an operational deployment force (ODF) the responsibility for which fell to 3ʳᵈ Brigade. This meant the brigade was generally better equipped, trained, and prepared for rapid deployment than the army's other two combat brigades. Units within the brigade were considered on-line or off-line with respect to their preparedness to deploy. When notified for deployment to Somalia, most of the soldiers and officers of 1 RAR had just left on Christmas leave, many leaving the Townsville area. This period is traditionally one of significantly reduced tempo for the ADF, but always with an expectation that members could be recalled from leave. Such recall was not unusual as a consequence of the need to support civil authorities in dealing with natural disasters like cyclones, flooding, and bushfires during this time frame (as is evident in our chapter by Chris Field).

1 RAR was reinforced for the deployment, thus its designation as 1 RAR Group. In addition to the six organic companies, 1 RAR was augmented by an armored personnel carrier (APC) squadron equipped with M113 armored vehicles, a civil military operations team (CMOT, centered on the battery commander and four forward observer parties from 1 RAR's direct support artillery battery), an engineer troop (equivalent of a platoon), and a logistic support company. The APC squadron and engineer troop generally provided direct support to the rifle companies during operations in Somalia.

The exercising and training activity levels in 1992 were unusually high, thereby serendipitously providing excellent preparation for

Operation Solace. Two of the largest exercises conducted were Joint Exercise Kangaroo and Joint Exercise Swift Eagle. The effort required to mount these two joint exercises meant they were normally conducted in alternate years. Fortuitously for 3rd Brigade, both took place in 1992.

The preparation for Somalia provided by exercises Kangaroo 92 and Swift Eagle 92 and associated preparatory training set 1 RAR up well for operations in Somalia. Two of the key exercise objectives for Kangaroo 1992 (K92) were (1) counter harassment in the north of Australia by a hypothetical enemy, and (2) define the extent to which the ADF would need to adapt its operational methods to be effective when confronting difficult physical environment conditions.[5] Swift Eagle 1992 (SE92) consisted of a mock NEO designed to further refine ADF tactics, techniques, and procedures (TTPs). Tasks included marshalling of civilians and dealing with unruly crowds, both of which would often be among the challenges we would confront in Somalia. Both events served as excellent preparation otherwise as well. The terrain was flat, the climate hot (and, when the season changed, wet). We therefore confronted other difficulties soon to be daily experiences, among them

- A large space-to-force ratio
- Few roads. Those available were of poor quality and largely impassable when wet
- Hostile forces demonstrating little intent to actively seek engagement with our better equipped and trained soldiers.

Further, developers had directed exercise civilians to behave in ways we later found to be very realistic, thereby providing us the opportunity to consider and develop effective TTPs.

Not only were these activities good preparation for soldier skills and company TTPs; they served as an excellent vehicle for the building of trust between Lieutenant Colonel Hurley and his company commanders. However, two of the four rifle company commanders that would deploy their subunits were not in command of those during this valuable preparation. There would thus be disparate levels of trust, familiarity, and expertise based on these training experiences so relevant to the application of mission command once in the Horn of Africa.

OPERATIONS IN THEATER: OPERATION SOLACE

Our assigned humanitarian relief sector [HRS Baidoa, which conformed to the Bai (Bay) region shown on Map 6-1] had a number of characteristics affecting the degree to which the CO employed mission command. The HRS was approximately 17,000 square kilometers (nearly 6,600 square miles) with Baidoa township situated in the center of the northern half of the area. The roads were generally in poor condition with the land for the most part flat and featureless aside from some hills and notably rugged escarpments. Bandits in the HRS had a very good human intelligence network and were always aware of our movements. We had no Australian or other dedicated aviation support but on specific operations were supported by helicopters of the US Army's 10th Aviation Brigade from Fort Drum, New York. Although our equipment included some of the earliest global positioning system (GPS) devices, maps were often wildly inaccurate, especially with respect to the locations of villages and unsealed roads that seemed to have moved many kilometers over time. We also had early modern night vision equipment to complement other systems dating from as far back as Australia's participation in the Vietnam conflict that we had borrowed from military museums in Australia. We also received a number of Australian strategic and some US assets with which 1 RAR had not previously operated. These included counterintelligence, psychological operations, and electronic warfare personnel and their accompanying equipment. We deployed for seventeen weeks with the expectation that we would not be replaced by other Australian forces.

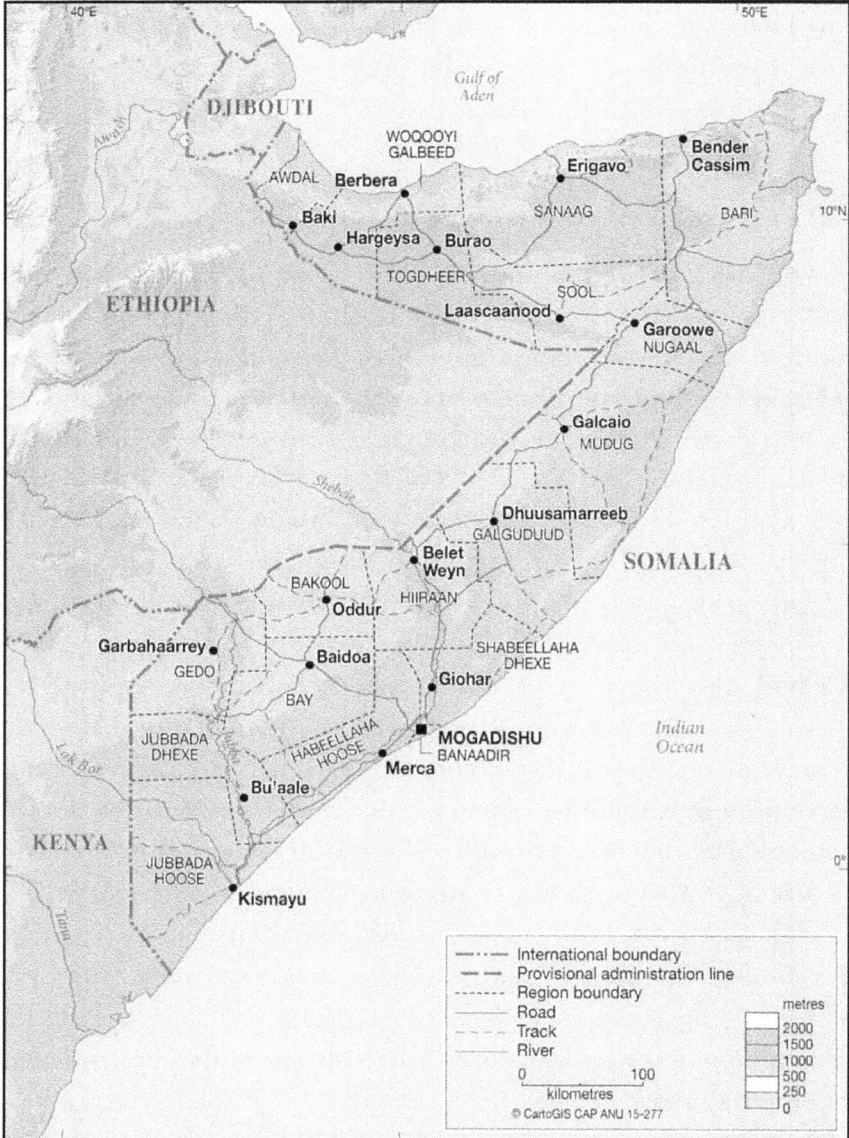

Map 6-1: Somalia[6]

These terrain, equipment, and other challenges impacted the speed with which decisions could be made and tempo of battalion operations. They also drove a need for highly decentralized, often virtually independent operations. Though beneficial in providing us key advantages such as superior night capabilities, they additionally meant we had some on-the-job learning to do quickly, especially if we were to achieve something noteworthy before returning home.

Hurley's two principal staff officers in support of his efforts to accomplish our assigned mission were me as OPSO (responsible for the detailed tasking and control of the operations of the 1 RAR Group) and the battery commander of the 107th Field Battery [chief CMOT] Dick Stanhope, who managed the interactions and relationship of our unit with the Somali elders and nongovernmental organizations (NGOs) who were the two stakeholders we were there to support. The two of us normally reported to the CO together to ensure LTCOL Hurley had a complete picture of his command's operations.

CMOT played a particularly important role in Somalia. The organization was officer heavy and each of its four teams had a vehicle with mounted communications. The teams generally operated in direct support of a rifle company as the main interface between local elders and NGO community and us.

The dispersed nature of our operations meant that the four rifle company commanders each had to have an intimate knowledge of the circumstances within which we were operating, the capabilities and TTPs of our own organizations, the battalion's mission, and CO's intent. A short description of the background of the four rifle company commanders demonstrates why the application of mission command by Hurley in Somalia could not be uniformly applied.

All four men were graduates of one of Australia's two officer-producing establishments, either the Officer Cadet School at Portsea or Royal Military College at Duntroon, having received their commissions roughly ten years before the battalion's deployment. All had undergone a series of professional military education and training experiences including the army's Grade 3 Staff Course (known at the time as the Junior Staff Course) and Infantry Centre Company Commander Course. Each had been in 1 RAR since Lieutenant Colonel Hurley assumed command, but—as noted— only two as officers commanding (OC) the companies they led in Somalia.

Major Doug Fraser had been OC A Company since his arrival in 1 RAR at the start of 1992. He arrived in the battalion at the same time as the battalion commander. Prior to his posting to 1 RAR, he had served as a platoon commander and OC Training Company in 5/7 RAR (Mechanised). Major Fraser had developed a solid relationship with Lieutenant Colonel Hurley commensurate with the two officers simultaneously establishing themselves in 1 RAR. Fraser had commanded A Company during K92 and SE92 in addition to taking the subunit to Operation Solace. He had a good understanding of infantry operations employing APCs from his 5/7 RAR experience and had a firm grasp of Lieutenant Colonel Hurley's expectations.

Major Jim Simpson had been the battalion adjutant throughout 1992. He was promoted to major and assumed command of B Company immediately prior to deployment. He had worked closely with Lieutenant Colonel Hurley as his principal personnel officer (S1) and as assistant operations officer (S31) during K92 and SE92. He had served in 6 RAR as a platoon commander and at the 1st Recruit Training Battalion as lieutenant and captain. Prior to his tenure as adjutant he had also been a reserve adjutant, a military observer with the United Nations Iran-Iraq Observer Group and immediately prior was an instructor at the Royal Military College, Duntroon. Major Simpson deployed to Somalia without experience as a rifle company commander but with a very sound and possibly the best appreciation of what Lieutenant Colonel Hurley expected of his subordinates and the way he expected them to undertake assigned tasks. This is particularly so given his working so closely with him as adjutant during which he was privy to Hurley's "thinking out loud when alone."

Major Mick Moon had been OC C Company for two years prior to the deployment to Somalia, having assumed that position on return from duty with the United Nations Truce Supervision Organization (UNTSO) in the Middle East. He had previously been a rifle company second-in-command and adjutant of 1 RAR before his deployment to UNTSO. He had therefore been in the battalion for four of the five years before deployment. Previous assignments included platoon commander with 5/7 RAR (Mechanised). Major Moon deployed with a very sound understanding of how 1 RAR operated and a good understanding of infantry operations employing APCs from his 5/7 RAR time. He also had a clear understanding of his relationship with the CO as a rifle company commander. Completing K92 and SE92 at the head of the company he commanded on Operation Solace, Moon was the most experienced rifle company commander in Somalia.

Major Ant Blumer replaced me as OC D Company two months before deployment, having been OC Administration Company for the ten months before. He was a platoon commander in 1 RAR for three years, subsequently undertaking French language training at the School of Languages (which proved very useful in Somalia) and then serving as aide de camp (ADC) to Australia's Chief of the General Staff (CGS). Blumer was the Australian exchange instructor at the Canadian Combat Training School for two years and was therefore well versed in the TTPs of the employment of attachments (especially snipers) and preparation of soldiers and officers for operations. Before returning to 1 RAR in January 1992, he took a year off from the army to hitchhike in Africa for most of 1991. As a result, he had extensive and contemporary experience regarding a number of African cultures. He had not commanded a rifle company during K92 or SE92 and as OC Administration Company he worked most closely with the battalion second-in-command. It was therefore natural that he had not developed the same degree of familiarity with Lieutenant Colonel Hurley prior to deployment. It is likely that his relationship with Hurley was the least well developed of the four company commanders.

It is clear that Lieutenant Colonel Hurley had good reason to exercise a different level of mission command based on his opportunities to form an appreciation of each man's skills. This was based on their varying career experiences, diverse tenure lengths as company commander (particularly for the two who commanded the same companies both during training in 1992 and Somalia), and familiarity based on both durations of acquaintanceships and positions held.

The 1 RAR Group mission statement delivered to commanders by Hurley before departing Australia was "1 RAR is to provide a secure environment for the distribution of humanitarian relief aid in HRS Baidoa." Hurley derived this mission from his evaluation of the intent behind United Nations Security Council Resolution 794 of December 3, 1992 that stated "action under Chapter VII of the Charter of the United Nations should be taken in order to establish a secure environment for humanitarian relief operations in Somalia as soon as possible."[7] Operation Solace operations centered on four primary, company-level tasks supporting the 1 RAR Group mission. These were:

- Secure Baidoa township, located several kilometers from the airfield from which the battalion based its operations. This task also involved protection of NGO base compounds in the township. This was the pre-eminent task.
- Protect NGO convoys and aid distribution.
- Secure of the Baidoa Airfield itself.
- Patrol-in-depth.

The first three company-level tasks were routinely undertaken on a nine-day cycle that allowed company commanders to cycle each of their three platoons through them for three-day periods. The three became relatively routine as time went by but remained vital. Each provided occasions when rifle company commanders needed to exercise disciplined initiative to solve new problems. This was most evident during early food distribution activities (when soldiers faced rioting by starving members of the local population) and on convoy protection duty when we were getting used to the modus operandi of the disparate NGOs.

Patrolling-in-depth provided company commanders the greatest freedom of action and need to capitalize on mission command. This fourth task involved patrols of either company or platoon strength, usually at distances exceeding one hundred kilometers from the group headquarters in Baidoa. Patrols were characterized by an inability for company commanders to communicate with either the battalion commander or his staff. Lieutenant Colonel Hurley thus granted his company commanders extensive leeway to develop understanding of their situation, solve problems, and establish and maintain security. Company commanders were allocated resources available at group level, virtually any not required for the other three tasks. Such assets included all or some of the following at various times: additional soft skin vehicles, an APC troop (platoon), a counterintelligence team, a CMOT detachment, an engineer splinter team [often with explosive ordinance destruction (EOD) capability and detection dogs], a sniper team, and a psychological operations team with interpreter.

In the early days, we very much stuck to fundamental principles until we became more familiar with our environment. We began operations by conducting all operations with more forces than were likely to be necessary until we were confident we could safely reduce the force levels. As time went by we found we could safely reduce the force levels and so increase the number of tasks and locations in the conduct of operations, which has been a characteristic of Australian operations in all conflicts. A good example is the convoy protections task in which we began with a company-level escort which within a matter of weeks was reduced on occasion to a section strength escort, albeit with additional area security being provided by other assets.

PRECONDITIONS FOR MISSION COMMAND

I address two preconditions in assessing the application of mission command in Somalia: the degree of trust between the commanders and the extent of common understanding regarding the battalion mission and commander's intent.

Trust involves characteristics of individual commanders' perceptions of fellow officers' reliability, honesty, and expertise. This trust is in part reliant on relationships developed thru common backgrounds and shared experiences. Trust, arguably the most important ingredient in applying mission command, is also dependent on operational context. A commander's ability, for example, will be assessed in terms of the nature of the task at hand and degree to which an officer's knowledge, skills, and past experience ready that individual to succeed. The level of trust shown is ultimately a function of these several variables.

Lieutenant Colonel Hurley's trust—and the trust in turn granted him by his company commanders—developed over time before and during the deployment to Somalia. It was underpinned by the four men's common professional military education. It further rested on the familiarity gained during the twelve months from when Hurley arrived in Townsville to assume command of 1 RAR, familiarity developed during training, in barracks, and on exercises.

Secondly, there must be a common understanding of mission and intent between commanders. These leaders need to be able to make decisions when the unexpected occurs or opportunities arise. An understanding of the senior commander's priorities and extent to which he condones variations from his instructions are essential if subordinates are to make decisions that best support the mission while acting within his intent. The degree to which a particular mission demands subordinate commanders employ initiative within an intent is also a function of a mission's character. Four rifle companies advancing along a battalion attack axis to seize an objective will require more specificity from their leader's intent regarding the nature of inter-company cooperation than for an Operation Solace-like undertaking during which there is little interaction between companies. Hurley acted to ensure common understanding of both his mission and intent from the day the deployment was announced.

The interplay between these two preconditions up and down the chain of command affected the degree to which mission command was applied on Operation Solace.

THE REALITY OF MISSION COMMAND ON OPERATION SOLACE

Reliability refers to a commander's and his subordinates' appreciation of the degree to which the expected response to orders and situations is repeatable. Is the battalion commander confident a given company commander's decisions are likely to be within expected tolerances? From a subordinate's perspective, is he as company commander confident in his understanding of the commander's expectations when making decisions when circumstances change? Can he depend on his senior to support him if an action taken in keeping with the battalion commander's intent has an unfortunate outcome?

There is no doubt that Hurley had confidence in the reliability of his four rifle company commanders. Each officer had ten years of army experience and had undergone very highly regarded individual and collective training regimes.[8] In addition, a posting as a company commander in a battalion of any type was a considerable achievement. Only the best officers in a cohort were selected as it was considered a prerequisite to later command at battalion level.

From the subordinates' perspectives, Moon and Fraser were the most confident in their understanding of Hurley's approach and his expectations of them as company commanders. Simpson and Blumer had had less opportunity to become familiar with those expectations when the operation commenced. Moon and Fraser, having been rifle company commanders throughout 1992, had served under Hurley as CO during K92 and SE92 in those roles. Simpson had a good feel for Hurley's general approach given his time as the battalion adjutant, but was less confident early in his rifle company commander role after having taken over B Company on promotion immediately before deployment. Blumer had less exposure to Hurley than Simpson and like him had not experienced Hurley's style of command from the perspective of a rifle company commander given his leading the Administration Company during K92 and SE92. It is unsurprising that Simpson and Blumer both sought time with Hurley during those early days in Somalia in order to better understand his expectations. Both sought to not only develop themselves, but also to benefit from participation in a battalion lessons learned process. They were keener to participate in collaborative learning, recognizing they would benefit from airing and gaining feedback from such exchanges. As the more experienced company commanders, Moon and Fraser received a greater share of the less-well-understood tasks during the opening weeks of Operation Solace and were expected to take the lead in TTP development.

The disparateness in Hurley's perceptions of reliability did not last for long. The school of hard knocks soon bought the four rifle company commanders much closer together in the CO's appreciation of their capabilities and judgment. As Simpson and Blumer became more familiar with the degree to which Hurley expected to be reassured regarding their reliability, they were tasked no differently than their more experienced peers and understood that interactions with the battalion commander could be less frequent. This maturation of relationships was particularly fortunate given the increasing demands placed on the battalion commander to spend more time addressing local elders and coordinating with NGO representatives. All four company commanders were comfortable in the level of trust Hurley afforded them by the end of the first month on operations.

Reliability was a vital, but not the only, underpinning of Hurley's trust in his four subordinates. Truth is another such foundation stone. Honesty is part of what a commander expects in terms of truth from his subordinates, but there is also the expectation to "tell it as it is," to be not only honest but to bring significant issues to the senior's attention no matter how unpleasant or otherwise and without presentation of facts to achieve a particular effect. One person's articulation of the truth can therefore be different from another's. What one considers significant another might not. Truth can also be measured by the degree to which a commander is prepared to "hear it as it is" whether or not the news is good.

Hurley made decisions and assigned resources based on his understanding of the progress toward mission accomplishment. His primary source of the battalion's status in this regard was his rifle company commanders. Sensing the pulse of Operation Solace via interpreting situation reports, battle updates, and occasional verbal debriefs from his sometimes distant company commanders was therefore an important task for the CO. Significant effort was required to obtain a balanced appreciation of the situation. This became more difficult as the time the CO had to devote to this pursuit diminished as he became more intimately embroiled in local affairs as the de facto military governor of the Bai Region.

Fortunately, interpretation of company commanders' reporting became easier as their expertise and familiarity with Hurley's expectations increased and the unexpected occurred less frequently.

While all four rifle company commanders were unquestionably truthful in their reporting, the relative importance of their part in the operation at a particular point in time was a significant factor in how they reported. Here again, context and experience was vital in the extent of trust the group commander would place on each officer's communications. Of the four tasks, patrolling-in-depth was least repetitious. Experiences differed both within company experiences and each time a patrol was undertaken by another company. It was resultantly patrol reports that demanded the most effort in cross-levelling the disparate emphasis company commanders placed in their accounts. The size of the battalion group's area of operations ensured there were never enough resources allocated to the company responsible for this task. What was allocated often came down to what could be spared from the other three tasks. The varied nature of patrols and distance from headquarters made it very difficult to forecast what resources a company would need for pending patrols. Over time, the CO's refined ability to understand what constituted "ground truth" for each rifle company commander and his improving understanding of the demands of specific locations and tasks meant decisions and resourcing became routine even for patrolling activities. Familiarity with his four leaders, trust in the validity and frankness of their reporting, and those subordinates' ever increasing expertise and understanding of their commander's intentions reduced the impact of distance and poor communications.

Ability—or expertise in the context of our broader consideration of mission command—is a combination of what is understood as KSA (knowledge, skills, and attitude) in Australian military circles. Ability is assessed against the nature of the tasks expected to be undertaken. A subordinate who is acknowledged as being able to execute a task well in one setting should not be expected to do so with the same level of ability in another if he or she does not have the same level of expertise in terms of those circumstances.

The significant problem for Hurley was that the expertise required to be successful in prosecuting the likely combination of tasks inherent in HRS Baidoa operations was initially not well understood. Professional education and training in the army had by and large focused on conventional operations. Fortunately, K92 and SE92 had helped in developing skills that later proved relevant in Somalia. Hurley's initial problem as he contemplated what manner of mission command to exercise with each of his four company commanders was to establish which of them had expertise best suited for the types of tasks we were conducting and how to provide priorities that would both abet mission accomplishment while also acting as training given the demands of Operation Solace's new challenges.

Early on in the deployment, Hurley attempted to tackle the uncertain and complex environment by managing the tasking of the four rifle company commanders, best utilizing what he perceived to be their strengths. Hurley had a very good appreciation of Moon's ability as the longest standing rifle company commander in the Battalion. In Townsville, the CO often turned to him when the unit needed training standards and field operations benchmarks. On operations, Moon was more heavily relied upon to guide how the battalion might conduct the protection of convoys and food distribution, and consequently picked up the lion's share of tasking early on. The latter of these was the task with which we were the most unfamiliar. Hurley rightly perceived that Simpson's strengths were somewhat different. He was well aware of Simpson's professionalism, and recognized his ability to operate effectively in the headquarters; however, Simpson had less practical experience as a company commander when operations involved some specific task types. For example, he had not had the opportunity to conduct an airmobile operation. Yet there were occasions when Hurley selected Simpson's company for a task as he would likely take more specific guidance without feeling like he was being over-directed. Despite his lack of experience with such operations, Hurley directed that he conduct the first airmobile operation of Operation Solace, which involved employment of lift helicopters from 10th Aviation Brigade. Fraser had the greatest degree of empathy with the CO given the background they shared in building relationships and reputation as they arrived in the unit together. In the early days, Fraser was expected to be the most flexible in changing the details of a tasking at short notice, particularly in defense of the airfield at Baidoa, as we learnt more about the requirement. Blumer had the most welcoming approach to the numerous and disparate attachments at our disposal. He was keen to learn-by-doing and debrief the results regarding utilization of these valuable assets for the benefit of others in the command. All four company commanders exhibited great propensity to learn quickly. The informal transmission of lessons between them meant that within a few weeks little thought was given to company commander proclivities and more to company tempo versus soldier rest and tasking variety to prevent monotony.

A significant issue arose in the execution of the 1 RAR Group mission as a result of a contrast in metrics used by CMOT and the rifle companies to gage when the group had achieved secure environment status. This conflicting assessment of the mission was the source of considerable discussion at each meeting of field grade officers whenever more than two got together.

The measures of success employed by 1 RAR and the Australian Army, both of which were in 1992 little experienced with the type of operations the group was undertaking, were oriented on decreasing enemy activity and number of enemy casualties. Experience gained as Operation Solace progressed (augmented by a specific threat to a NGO at one stage) led to recognition that success was NGO rather than military force based. Failure would be the NGOs deciding they could not provide humanitarian relief under what we considered a secure environment. NGO departure would make our presence superfluous. Success therefore meant that the NGOs were able to continue their activities because the environment was secure *from their perspective*. This reality was first understood by the CMOT, later the CO, and eventually by me and the rifle company commanders. The challenge then became one of moving from what we had considered a secure environment to that as perceived by the NGOs.

It was clear to us that there were not enough Australian soldiers to replace all of the guards that the NGOs believed were needed to secure their operations. In Chief CMOT's view, therefore, he was compelled to issue weapons licenses to those designated as guards for NGOs (local civilians otherwise being prohibited from carrying weapons as mandated by UNITAF). However, the soldiers of the 1 RAR Group quickly discovered that these guards-by-day were the bandits by night, a fact that was not well understood early on. The situation came to a head when one night a C Company patrol came under fire as they passed a building. After, assaulting the structure, the soldiers found Somalis inside holding up their licenses as justification for their possessing weapons. It was obvious that CMOT licensing had the unintended consequence of putting Australian lives at risk.

This problem was something of a mission command conundrum. It did not come about due to a poorly worded mission or misunderstood commander's intent. Both the leaders of the rifle companies and CMOT had been conducting operations in a manner they thought was in keeping with Lieutenant Colonel Hurley's mission statement and supporting intent of providing a secure environment for NGOs. CMOT's licensing NGO guards was an agreed practice, the consequences of which were not fully understood, likely due to inadequate intelligence. As noted, however, the practice turned out to be dangerous for patrolling rifle company soldiers. In hindsight, given that we eventually occupied the NGO base compounds, we should have come up with a plan that made the use of Somali guards nugatory. That might even have involved moving some or all of them onto the airfield at Baidoa. For the purposes of us assessing all the implications for the mission, we needed not to assume the intelligence picture was the complete picture.

There is sometimes a misconception that a commander's intent statement in a mission order is all that is required to ensure subordinate actions will be in alignment. The complexity of contemporary operations ensures this is never the case. The greater a subordinate's understanding of his commander's beliefs and values, the more likely that subordinate will act as desired. Actions speak louder than words. A CO's reaction to an event, the manner in which he or she deals with an incident, and the degree to which a subordinate receives support when things go wrong all send strong messages that influence a junior leader's appreciation of their commander's declared intent and the trust bestowed.

Many of us in 1 RAR had a good feel for the CO's implicit intent by the time we deployed, particularly after the previous year's cycle of intense exercises and training. The aforementioned differences in familiarity meant some company commanders had a better grasp of how Hurley was likely to expect things to be done than others.

Hurley's intent for Operation Solace was clear. It elaborated on his understanding of the mission, but it was the feel for how he expected his commanders to perform that most supported his effective employment of mission command. It was this combination that ensured rifle company commanders' initiatives were disciplined and bound by more than the words of the group commander's mission orders.

Most importantly, Hurley demonstrated a great faith in the Australian professional education, training, and career management systems. He trusted that the system identified officers most capable of successfully meeting the demands of higher command. He also had a high regard for the professionalism of the Australian soldier. We underwent significant training regarding rules of engagement throughout 1992, particularly in connection with the NEO training and in the lead up to the December deployment. His reaction to a US marine incident in Mogadishu we watched on television before we deployed demonstrated Hurley's faith in our judgment to understand ROE holistically rather than be hostage to specific wording. In a 1995 interview, Hurley described our soldiers' approach to operations as "bigger than the ROE." Addressing lessons from Somalia in that interview, he pointed to "a number of occasions where [our soldiers] could quite lawfully have shot someone...but they were able to judge that the situation didn't require that response."[9]

Hurley additionally demonstrated significant empathy for the local population and a strong desire to be sure we were doing what was necessary to gain their support. He was a student of history and we quickly turned to the Templer model employed in defeating Malayan guerrilla insurgents between 1952 and 1954.[10] Implicit elements of Hurley's intent included the need for subordinates to understand that our operation was intelligence led and that local elders were to be treated with appropriate respect.

Hurley recognized that intelligence collection and analysis would be particularly critical if we were to secure an environment in which the enemy did not wish to take us on directly. We worked hard to optimize the intelligence we had at our disposal and integrate it with operations. We had significant intelligence assets and raw information at our disposal, capabilities that included the previously mentioned human intelligence, electronic warfare, and psychological operations teams. The battalion had not trained or exercised with most of these intelligence resources, nor was it experienced in handling the volume of information pouring into the command post. The massive input from our companies and that inherent in the broader picture provided by our higher headquarters at 10th Mountain Division (an organization unfamiliar with dealing directly with units without a brigade HQ between them) added to the challenge, a challenge we did not handle as effectively as we might have. The group intelligence officer (S2) at the time was an infantry captain rather than an intelligence professional. The Australian Army would in later years designate intelligence cell positions as requiring professional intelligence personnel.

Our focus on the population was very much in keeping with Templer's "hearts and minds of the people" approach. The task of dealing with the local population was initially left to CMOT. This was partly because we did not have a good feel for the complexity of the human terrain; it also allowed the CO to focus on establishing a successful operational pattern. However, his attention was increasingly drawn to the local elders who wanted to deal with the "boss." Ever gaining greater appreciation for his rifle company commanders' capabilities, Hurley gave them even greater freedom of action. The CO believed in ensuring he dedicated time to that which only he could do effectively.

CONCLUSION

Mission command as practiced by Lieutenant Colonel David Hurley and his four rifle company commanders contributed significantly to the success of Operation Solace. Key contributing factors included his depth of trust in those men and the decentralized nature of his tasking.

The degree of trust Lieutenant Colonel Hurley granted his rifle company commanders was born of his faith in the army's selection system and extent he worked with them over the twelve months preceding deployment. Under Hurley, the battalion had patterned its training on the enemy portrayed in the government White Paper

"Defence of Australia 1987," released less than five years earlier. Two major exercises in 1992 had prepared us well for the type of low-level operations both emphasized in that white paper and realized in Somalia, particularly regarding the demands of working closely with civilians and managing large crowds. The terrain and enemy were likewise similar to those confronting us during that year's training in northern Australia. We were fortunate in that the army was already contemplating how best to deal with problems like how to deal with widely dispersed operations when both road networks and communications were poor. The battalion further benefited in being one of the army's best trained, equipped, and resourced by virtue of its having been part of the ODF since the late 1970s. These factors combined to ensure leaders' confidence in each others' reliability and expertise.

While the consequences of a less well understood and specific intent were not as great during Operation Solace as is likely to have been the case for conventional operations, an implicit intent taking into account the experiences of each subordinate over the previous twelve months was paramount. Rifle company commanders understanding of Hurley's intent was built as much on their understanding of their group commander based on those twelve months of preparation as on the specific intent for Somalia.

The operation conducted by 1 RAR Group was for the most part conducted as four separate company-level tasks conducted as four decentralized operations. The quartet of tasks provided a means of imposing desirable structural partitions between the operations of the rifle companies. Our training and operational experiences had influenced our doctrine; such a division of labor was considered the most effective way to organize the force for a mission with the conditions we faced in Somalia. The rifle company commander assigned the patrolling-in-depth task in particular enjoyed considerable freedom of action within the resources allocated him and boundaries assigned.

Lieutenant Colonel Hurley confronted a situation with significant unknowns as 1 RAR Group arrived in Somalia in 1993, but he trusted his rifle company commanders, albeit in somewhat varying ways and degrees. He reduced force-to-task levels as we became more comfortable with our understanding of those necessary to maintain a suitable level of security. Hurley likewise increasingly diverted his energy toward grappling with the problems only he could deal with: coordinating with Somali elders and NGO leaders. After recognizing that mission success rested first and foremost on supporting NGO delivery of aid to the civilian local population, there was a serious risk of mission failure if the perception among NGOs and Somali populace was that the bandits from which we were protecting them would prevail. The group commander thus came to realize that there was a lesser risk of operational failure were a rifle company commander to fail in some way while the Australians maintained that ability to support.

The successful conduct of 1 RAR Group mission in Somalia was in large part due to the employment of mission command. Without the trust established between the CO and his subordinate rifle company commanders and his specific taskings and adapted guidance, the battalion commander would not have been able to devote the time necessary to dealing with the keys to that success: Somali elders and NGO leadership.

ENDNOTES

1   Major James Simpson, *In Retrospect: An Oral History on an Operation Other than War,* unpublished paper submitted to Australian Army Command and Staff College, copy provided to author by Major [now Brigadier (Australian Army, retired)] Simpson, Fort Queenscliff, 1995, Glossary-iii.

2   Chief of Joint Operations, *Executive Series ADDP 00.1 Command and Control* (Canberra: Commonwealth of Australia, 2009), 2-8.

3   "United Nations Security Council Resolution 794 (1992)," December 3, 1992, http://www.un.org/en/ga/search/view_doc. asp?symbol=S/RES/794(1992) (accessed September 29, 2015).

4   Defence of Australia 1987, Department of Defence, Canberra: Australian Government Publishing Service, March 1987, 59-63, http://www.defence.gov.au/Publications/wpaper1987.pdf (accessed December 20, 2015).

5   The Parliament of the Commonwealth of Australia, *Report of the Visit of the Defence Sub-Committee of the Joint Committee of Foreign Affairs and Trade to Exercise Kangaroo 92 19-21 Mar 1992,* Canberra: House of Representatives Printing Section, May 1992.

6   Map adapted from Australian National University CartoGIS, item no. 15-277a_Somalia_colour-01.pdf.

7   "United Nations Security Council Resolution 794."

8   Ten years later I was appointed Chief of Staff, HQ Training Command. In all of my interactions with US, British, and Canadian counterparts, the overwhelming perception of

our individual training system was exceptionally positive and many a visitor was sent specifically to see how Australia managed to achieve such a high standard. On combat operations subsequently, particularly in multinational headquarters, I was always impressed to find that our officers were highly sought-after for command and staff positions by Australia's allies.

9 Simpson, *In Retrospect*, B-4 (Answer to Question 6).

10 Field Marshal Sir Gerald Walter Robert Templer KG, GCB, GCMG, KBE, DSO was a senior officer in the British Army. He is best known for his role in defeating guerrilla rebels in Malaya between 1952 and 1954.

7

AUSTRALIA'S APPROACH TO MISSION COMMAND DURING THE EAST TIMOR INTERVENTION 1999

Dr. John Blaxland

Austraila's involvement in the liberation of East Timor in 1999 was the most decisive demonstration of Australian influence in the region since World War II and the country's largest military contribution since the Vietnam War. Australian leadership was instrumental in the events that led to the birth of the region's newest nation. So to what extent was the conduct of this operation a demonstration of Australia's approach to mission command? There are a number of angles worth considering given the level and scope of Australia's involvement in this operation. The practice of assigning subordinate commanders with missions without specifying how the mission was to be achieved was a hallmark of the Australian-led intervention in a number of ways. This chapter provides some context to the mission before examining the command and control arrangements which set the scene and the tone for the exercise of mission command. The chapter then examines the exercise of mission command with Australian and coalition forces at various instances during the conduct of the Australian-led intervention. But first to some context.

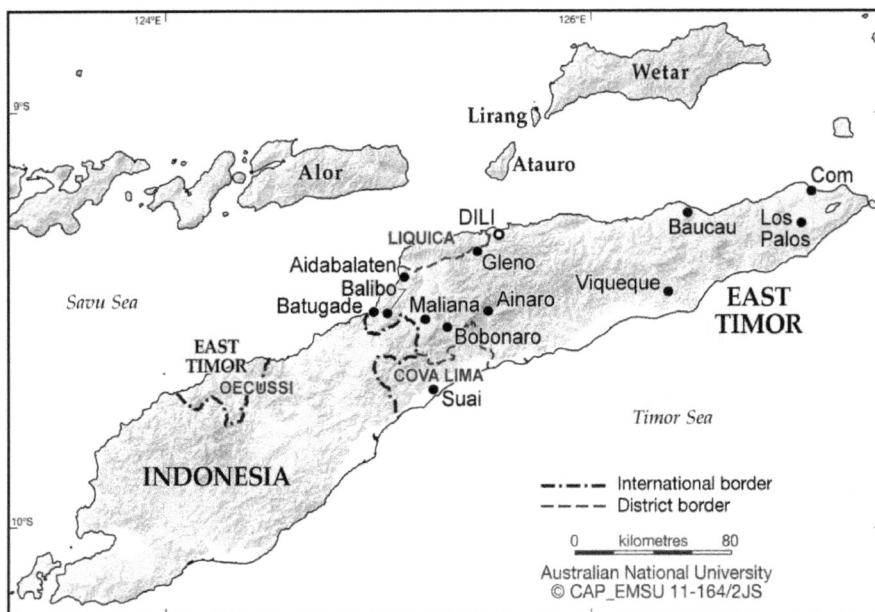

Map 7-1: East Timor/Timor Leste[1]

LEAD UP TO INTERVENTION

President Suharto had been in power since overthrowing his Communist-sympathizing predecessor in the mid-1960s. Concerned about the prospect of a mini-Cuba on their doorstep, Indonesia annexed militarily the neighboring previously Portuguese territory of East Timor in 1975 shortly after a local leftist Timorese group declared independence. In the years that followed, Australian-Indonesian relations were troubled, particularly over the issue of harsh Indonesian handling of the province – including the murder of five Australian journalists who were caught witnessing the annexation. Ostensibly non-aligned, Indonesia established strong links with the United States, overseeing unprecedented economic growth during the Cold War and beyond. The Asian Financial Crisis of 1997, however, saw a number of countries in Southeast Asia experience exceptional economic turmoil. In the case of

Indonesia, it also led to the demise of President Suharto in May 1998 following riots in the capital Jakarta. He was replaced by his deputy, the mercurial German-trained technocrat, B.J. Habibie.

Habibie's rise to power marked a significant inflection point for Indonesia's handling of the province of East Timor. With the new president in Indonesia, Australian Prime Minister John Howard saw an opportunity to help broker an autonomy measure for East Timor akin to the 1988 Matignon Accord brokered between France and the people of the French Pacific Ocean territory of New Caledonia. The proposal backfired. Habibie insisted on precipitating a final determination promptly by calling for a plebiscite. The United Nations Assistance Mission in East Timor (UNAMET) was established under UN Security Council Resolution (UNSCR) 1246/99. Participants in the mission, including some international police contingents and UN military observers, were sent unarmed to East Timor in June 1999 tasked to organize, conduct, and supervise a referendum to allow the people to choose between autonomy within Indonesia or independence.[2]

Following the overwhelming vote for independence on 30 August 1999, the militias raised in an effort to ensure a favorable outcome for Indonesia began attempting to disrupt the election result. International contingents eventually either left the country or sought refuge in the Australian diplomatic compound in Dili. This situation triggered an Australian-led evacuation operation from September 6-14 using mostly Royal Australian Air Force C-130 Hercules aircraft to evacuate hundreds fleeing militia terror. But a far larger operation was about to present itself to the international community, particularly Australia. By September 12ᵗʰ, under intense international pressure, Indonesia agreed to accept a UN-mandated international force to restore order.

Facing domestic outrage over events in East Timor, the Australian government agreed to lead the multinational force. This was a significant departure from the more constrained approach to expeditionary commitments that Australian governments had followed since the end of Australia's military commitment to the Vietnam War.[3] Australia's experience in monitoring a peace agreement in Bougainville from late 1997 to 1999 had helped prime the Australian Defence Force (ADF) for this event, but that was on a far smaller scale with only a handful of countries participating and involving only unarmed monitors.[4]

Commencing on September 20th, INTERFET deployed with contingents from twenty-two nations initially including Australia, Brazil, Britain, Canada, France, Italy, New Zealand, the Philippines, Thailand, and the United States. Others followed afterwards.

One could be forgiven for thinking the intervention was not such a big deal; after all, while many suffered and died in East Timor prior to the arrival of INTERFET, the intervention itself went remarkably casualty-free. The insertion went smoothly, devoid of the major clashes between troops forming part of INTERFET, militias, and even the Indonesian National Military (*Tentara Nasional Indonesia*, or *TNI*) that some predicted might occur. In fact, while studiously seeking to avoid such an outcome, planning in the lead up to the deployment had anticipated the prospect of possibly hundreds of coalition casualties. How this was avoided has much to do with how command was exercised at a number of levels. Before considering the exercise of mission command on operations in East Timor in detail, it is worthwhile outlining the command and control arrangements set in place for the operation to help understand how they set the tone for the exercise of mission command in the field.

Oversight, Command and Control Arrangements

The Australian government's management of the East Timor crisis reflected developments with the structures, processes, and norms that had been emerging in the national security realm in

Australia's capital, Canberra, since the mid-1990s. As a British Westminster-styled parliamentary democracy with elements of the US federal bicameral system incorporated, Australia's system of government is unique. David Connery has observed that foremost among the elements of the governmental system engaged in the East Timor crisis was "a committee system that brought advice together with decision, and created a dominant prime minister and closed national security executive." He further noted that these arrangements created an imperative to coordinate effort across the various arms of government while keeping in touch with important external parties. In turn, experience with these mechanisms of government led to an emphasis on "whole of government" operations in the national security field.[5]

Connery observed that the first lesson learned involved the prime minister's role as the "minister for national security." This role, he argues, "comes to the fore whenever new directions are needed in foreign policy, whenever military force is needed or whenever there is a major disaster." To Connery, "this is the prime minister as sense-maker to the Australian people, statesmen to outsiders, and key decision maker and coordinator inside government." He further observed, "prime ministers tend to embrace their national security role and use it to pursue the national interest while they extract the maximum political advantage."[6]

For Howard to exercise this kind of role required reliance on extensive oversight mechanisms. In particular, the management of the East Timor crisis was enabled by establishing a number of mission-specific bodies. The three key organizations were the East Timor Policy Unit in Australian Defence Headquarters (ADHQ), INTERFET Branch within Strategic Command under ADHQ (commanded by Major General Mike Keating), and the Department of Foreign Affairs and Trade (DFAT) East Timor Crisis Centre. In addition, an inter-departmental committee chaired by a representative from the Department of Prime Minister and Cabinet was implemented to coordinate policy and whole-of-government actions. These ad-hoc structures and processes promoted flexibility and more responsive decision-making. They dovetailed with the established mechanisms associated within higher level government machinery, especially the National Security Committee of Cabinet chaired by the prime minister where final decisions on Australian military commitments are made. Confident that the best inter-departmental advice had been garnered, the prime minister could act decisively. In turn, this clarity of decision, authority, and responsibility helped provide the critical support necessary for the military chain of command planning and conduct of operations. This higher level support was important in allowing the freedom of action necessary to exercise mission command with deployed force elements.

This domestic clarity was reinforced by UN Security Council Resolution 1264/99 that provided a clear mandate to restore peace and security in East Timor; to protect and support the UN in carrying out its task; and, within force capabilities, to facilitate humanitarian assistance operations. It was enacted under Chapter VII of the Charter of the United Nations, giving INTERFET the freedom to act decisively, using lethal force if necessary and enabling the force to take the initiative away from hostile groups. Australia's Brigadier Mike Smith, who was seconded to work at UN headquarters in New York with the Department of Peace Keeping Operations (UN DPKO), played an instrumental role alongside his DFAT counterparts in ensuring the wording of UNSCR 1264/99 was sufficiently robust to enable the mission to be undertaken successfully.[7] Smith worked closely with Keating in Strategic Command to ensure this outcome. Keating and Smith knew each other well and trusted each other. This trust and understanding reflected not only their common roots as graduates of Australia's principal officer training institution, the Royal Military College Duntroon, but also shared experiences as infantry commanders and successive commanders of Australia's Operational Deployment Force based around the army's 3rd Brigade. For them the exercise of mission command was implicit in their day-to-day actions and consultations.

Once the international mandate was ensured, the Australian Chief of Defence Force (CDF), Admiral Chris Barrie, appointed the commander of Australia's Deployable Joint Force Headquarters (DJFHQ), Major General Peter Cosgrove, as Commander of the International Force East Timor. Cosgrove was placed under command of the CDF and reported through him to the government of Australia and United Nations with day-to-day communication also maintained between Keating's Strategic Command and Cosgrove.

Figure 7-1: Admiral Chris Barrie, Prime Minister John Howard, and Major General Peter Cosgrove, East Timor, 1999[8]

Given the wide span of responsibilities Cosgrove faced (unprecedented for an Australian commander), a Commander of Australian Contingent (COMASC) was appointed to exercise national command of Australian forces assigned – about 7500 ADF personnel out of an international force of 11,500. Initially this was Brigadier Mark Evans, commander of the main Australian land force assigned, 3 Brigade.[9]

A robust and unified command structure, coupled with the deployment of balanced forces from all three armed services alongside coalition forces, was instrumental to success in East Timor. Formation-level collective training and the high level of individual soldier skills were additional important factors. Just a few months before the operation, a command post exercise was conducted based around the Deployable Joint Force Headquarters in Brisbane using a scenario not unlike East Timor. That exercise involved members of the armies of the ABCA countries: America, Britain, Canada and Australia. It helped to ensure that techniques and procedures were well understood and practiced in advance.[10] This practice of meeting together, benchmarking each other's performance, and testing limits of capabilities on such exercises would prove significant in East Timor. Armed with the knowledge of the experience and outcomes from the exercise, Cosgrove and his staff could make judgments about task assignments with a level of confidence critical to effective mission command.[11]

SIGNIFICANCE OF COSGROVE'S APPROACH

In an operational environment short of war like in East Timor in 1999, Cosgrove operated with what could be described as a maneuverist mindset attuned to the information era. Combat actions were subordinated to other goals through a focus on messaging and influence to obtain outcomes with minimal reliance on the use of lethal force. To do this, Cosgrove made use of the gathered international media to shape and influence views of significant participants in the unfolding drama.[12] Being largely favorably disposed, the media effectively value-added to the traditional combat power of the military forces deployed. Cosgrove was an excellent communicator with the media, helping to shape and influence events in a non-kinetic way to contain physical and political collateral damage. He masterfully employed his forces with restraint and resolve and ensured that the messages passed through the media reinforced his objectives.

Given the challenges, reviews, and cutbacks experienced in the years preceding the operation, Cosgrove justifiably observed that the mission proceeded smoothly despite the ADF's shortcomings, saying, "We were lucky and we were good." Cosgrove stressed the two sides to this coin. Often enough luck comes from hard work built on mutual trust and respect of common professional standards that made the force "good." For Cosgrove, this worked both up and down the chain of command and responsibility. He was trusted by Keating and Barrie, for instance. Barrie was in turn trusted by Howard, and Barrie trusted Keating. Admiral Barrie in fact credits the advice and support provided by Keating from May 1999 onwards, recalling

> Mick Keating and I had been colleagues in the Australian Staff College Course of 1977 at Queenscliff, so we already had some knowledge of each other's work style. Mick's sage advice about the conduct of operations and his ability to manage the command relationships between [key senior officers was] the product of extensive experience and wisdom in the profession of arms involving all three Australian services and the contributions of our coalition partners where necessary. His invaluable insights and willingness to talk to people at all levels in the lead up to and during Operation INTERFET provided a sound basis for our overall success and complemented the manifest skills of the other deployed commanders.[13]

In a manner akin to that of Barrie and Keating, Cosgrove demonstrated his own understanding of the essence of clear direction and trust of subordinates delegated to perform specific tasks. Cosgrove focused on the strategic imperatives inherent in Australia's participation in INTERFET. First and foremost was to help restore peace and security to East Timor as quickly as possible. Second, to do so in a way that enhanced our relationship with the Republic of Indonesia. And finally, as the international leader of the INTERFET coalition, to assist our partner nations to optimize their participation in the INTERFET mission. He noted:

> Some may wonder at a certain lack of subtlety in this articulation of our strategic aims. Consider, though, whether further nuancing of these simple statements would have been useful or perhaps distracting in the framing of operational activity. In the very complex environment that developed on that day fifteen years ago (and arguably in the frantic week beforehand) it was vital to have a core set of requirements to test whether we were on the right track, and I used these Australian strategic imperatives as my mantra in my dealings with our coalition partners. These partners were in furious agreement about the "restoration of peace and security" leg, very enthusiastic about "optimizing their participation in the INTERFET mission" part, but sometimes unclear about the priority and the modalities of nurturing the relationship with Indonesia. It was with some irony that we came to appreciate that Australians were more expert in this regard than all but a few of our coalition colleagues.[14]

Cosgrove understood that for the principal task to be achieved with the minimum of bloodshed and collateral political damage to the bilateral Australia-Indonesia relationship, he needed to work on trust between himself and his subordinate contingent commanders. Even more importantly, he also had to do so with his counterpart, the Indonesian Army's Major General Kiki Syahnakri. Cosgrove observed that

> There was significant disenchantment between Australia and Indonesia in September 1999, and no matter the reasons or the validity of that disenchantment, it was very important for the future of our relationship that we worked very hard and keenly to focus on avoiding friction and exhibiting goodwill. The best place to do this was literally "from the ground up" — our Australian attachés started this during the evacuation operations out of East Timor earlier in September.
> General Kiki and I embarked on this process when we met in Dili for the first time. We met there virtually daily thereafter and, when I was being burnt in effigy in Jakarta, in Dili he and I would be discussing issues of local security and military de-confliction over coffee in a professional and amiable way. [15]

Given the politically sensitive and potential enduring damage to the bilateral relationship with Indonesia that could be caused by overplaying reliance on lethal force, Cosgrove and his Australian brigade commander, Evans, appreciated the need for a nimble-minded approach. They realized that effective information operations could back up the psychological effects of nonlethal force and directed their staffs to plan accordingly. The two men also recognized that they had to keep collateral damage to a minimum in accordance with the rules of engagement. That was critically important for the sake of Australia's long-term political relationship with its neighbors. It was also necessary in order to maintain cohesion within INTERFET.[16]

> Cosgrove understood that fundamentally, he had to command and support Australian and other international forces in the most professional way.

Plans and orders had to pass the litmus tests of being useful to the overall mission, viable and logical for the forces involved, even-handed and invariably respectful and sensitive. The end state for all partners had to be a conviction that the overall mission and their particular part in it was worthwhile and that Australia's leadership role was both useful and, if ever necessary again in the future, repeatable.[17]

Reflecting afterwards, Cosgrove observed

While personally I can hope and believe that this may have been the case at the end of INTERFET, one thing I am very clear about is that Australia can hold its head high on the professional standards, energy, and effectiveness of its stewardship of the INTERFET deployed forces. Inevitably there were some bumps in the road—sometimes through faulty communication and sometimes through the misreading of nuanced messages between partners—but I am very clear that overall we did this part of the job quite well.[18]

Cosgrove further understood that in keeping international contingents compliant, and happily so, considerable energy had to be expended in relationship management. He explained it this way:

After we had sorted out the planning and operational routines and appropriate deployments for national contingents and some individuals, the coalition focus turned to endlessly honing communications and equally endlessly, "socializing" the force through visitation, consultation, and even a heavy program of cultural events. It worked, and I would do it again. [19]

Cosgrove recognized both the United Nations and NGOs as fundamental and important stakeholders. In fact, one of the reasons INTERFET existed was to facilitate the speedy return by the United Nations and its early control and coordination of civil operations in the new nation. He recalled:

> If I met General Kiki daily, then I met Ian Martin, Special Representative of the Secretary-General, and later Sergio Viera De Mello, at least daily and often more as we progressed from a military-led security operation to a UN-led rehabilitation operation. [20]

With regard to nongovernment organizations (NGOs), Cosgrove noted he had "a steep learning curve in relation to their philosophy, structure, and operational imperatives when dealing with their needs and representations in the first few days of the INTERFET mission." He acknowledged that NGOs operate to provide humanitarian support in the areas of food relief, shelter, medical support, and sometimes education and social development involving such initiatives as employment programs and specialized programs for the empowerment of women. He recognized that some NGOs also span a number of different programs, but "naturally all are single-minded in the provision of their particular set of capabilities to the needy in the host community."[21]

Cosgrove identified another important factor to remember: "There will always be a dynamic between military security operations and the high priority political, social, and infrastructure operations of the civil authorities (where some feel that any nearby soldier should quietly pack up and leave the vicinity when civil authority representatives come into view)." Regarding all of this, Cosgrove observed,

> The closest, most constructive, professional, and hopefully personal relationship ought to exist between the military commander and the leader of the civilian authority. Sergio and I

enjoyed this and I had the greatest regard for him. It meant that all the inevitable moaning and groaning among subordinates could be addressed and resolved amicably and quickly.[22]

This focus on management of key relationships was fundamental to the effective exercise of mission command in East Timor. A less attuned commander may well have allowed mistrust and misunderstanding to fester, thus undermining the critical element of trust implicit in mission command.

TACTICAL PROFICIENCY AND THE EXERCISE OF MISSION COMMAND

While the national governance and oversight arrangements played their part in ensuring the space for effective mission command, Cosgrove's leadership certainly played a pivotal role in ensuring the mission's success. But in doing so, he also built on the foundations of a highly professional and reliable force that was able to execute tactical actions directly in line with his senior commander's intent: that is, restoring peace and security while enhancing the relationship with Indonesia and assisting participating partner nations' forces. This section examines a few examples of tactical actions that demonstrate higher commanders' trust and subordinates' clear understanding of their commander's intent.

In the days prior to the arrival of INTERFET, Australia's defense attaché to Indonesia, Brigadier Jim Molan – supported by elements of the Special Air Service (SAS) under Major Jim McMahon – evacuated East Timor's Bishop Belo from Baucau (a town east of the Timorese capital of Dili). They did so after a tense standoff with thirty militiamen and forty Indonesian soldiers at the airfield.[23] The experience demonstrated the significant roles that military attachés with a clear understanding of the mission and purpose can play in support of such operations. The collaboration between

Molan and McMahon's team also meant that what could have turned out to be a disastrous end for Bishop Belo passed relatively uneventfully and unnoticed. McMahon recalled that his small team had a multitude of assets responding to his direction until they "cut over" (i.e., reverted to being commanded and controlled by higher headquarters) when the rest of the force arrived. Reflecting on this experience later, McMahon observed, "We shouldn't shy away from giving the commander on the ground all the assets he needs to get the job done, with assets above water [and] under water for periods of time also attached. We didn't just turn up on the day—there was interplay with a range of components."[24]

On September 20th the INTERFET Response Force, based around McMahon's SAS squadron, deployed to secure Dili's Comoro airfield. The 2nd Battalion Royal Australian (Infantry) Regiment (2 RAR) under the command of Lieutenant Colonel Mick Slater, the remainder of 3 RAR, and their combat support and service support elements from the brigade followed close behind. The deployment of the two battalions (with a field artillery battery operating as an infantry company as part of 3 RAR), supplemented by a reinforced British Gurkha company group from Brunei and Royal Australian Air Force airfield defense guards, enabled INTERFET to begin the process of restoring law and order. This involved conducting patrolling, establishing vehicle checkpoints, apprehending members of the militia, and disarming them. Sporadic gunfire continued and smoke still billowed from gutted buildings as patrols uncovered bodies, some of them mutilated, around the Dili area. This disparate group of combat elements worked together confident in the knowledge that each team had clear and unambiguous orders from the force commander and each component had complementary and unambiguous roles to play to ensure the mission's success.

With news of violence continuing in the eastern end of East Timor on September 22nd, A Company, 2 RAR conducted an airmobile operation to secure Baucau airfield to the east of Dili, deploying on 5 Aviation Regiment Black Hawk helicopters. This act demonstrated INTERFET's ability to move swiftly and decisively. With little guidance and short preparation time, the company operation nonetheless proceeded according to plan. Three days later the unit handed over the airfield to a Filipino infantry company and returned to Dili. The operation demonstrated the Australian and Filipino forces had a clear understanding of the commander's intent, proceeding to carry out their orders with minimal direct oversight.

In another instance, McMahon was tasked to deploy troops to the port of Com on East Timor's northeastern coast from which there had been reports of militia fighting. McMahon's force was supported by Black Hawk helicopters and the Royal Australian Naval ship HMAS *Adelaide* as well as a company of infantry to move on short notice in case there was a need for additional troops. Moving on foot at last light, McMahon's force identified about 2,500 East Timorese interspersed with militia guards. With a clear understanding of his commander's intent, McMahon gave the militia an ultimatum to surrender or face arrest. Later that night a group of twenty-four armed men sought to leave the compound on a truck; McMahon quickly deployed his forces to surround the truck, had his troops shine torches on the armed group, and called for them to drop their weapons. The shock of surprise led to their quick surrender. The raid to Com was an exceptional performance by McMahon and members of his force.[25] The raid also demonstrated the utility of highly trained soldiers able to deploy over distances to achieve assigned task against armed adversaries without having to resort to lethal force or seek clarification of their mission from higher headquarters before undertaking time-sensitive actions. Instances such as this and subsequent contacts that occurred during the first half of October demonstrated the effectiveness of Australian forces.

Ground force capabilities also were significantly enhanced by contributions from a number of countries. Troops from coalition nations came with their own professional skills and heritage, complementing the Australians with their unique strengths that contributed to the successful outcome of the mission. Cosgrove clearly understood the need to manage expectations and provide unambiguous guidance to these various national contingents. Exercise of mission command, one could argue, was less feasible with contingents lacking common standards or shared experiences upon which to build trust, mutual confidence, and respect. Indeed, it was no mean feat to enlist and then maintain the support from so many diverse countries with different cultures, languages, and religions. Meeting the demands that arose challenged the ADF's ability to supply interpreters and liaison officers for the various contingents. But the fact that the army had exercised with many of them in the past helped considerably in facilitating their integration into the coalition.

Thai troops, for instance, were familiar with techniques for winning hearts and minds given their long and successful fight against communists in remote parts of their country. Their battalion operated in the more benign eastern sectors of East Timor, away from the border with West Timor near to where the militia had regrouped on the Indonesian side. A company from 2RAR had only recently completed a combined exercise with the same battalion in Thailand. This experience undoubtedly helped add to the level of mutual confidence and trust.[26]

The New Zealand Defence Force was amongst the easiest of coalition partners for the Australians to work with. The commanding officer of the 1st Battalion Royal New Zealand Infantry Regiment (1 RNZIR), Lieutenant Colonel Kevin Burnett, observed that there was considerable value in 1 RNZIR having deployed as a battalion with the 3rd Brigade on exercise in northern Australia two years previously. That exercise was an opportunity to renew the close relationship with the 3rd Brigade and units with which 1 RNZIR had close ties since the Vietnam War. Burnett observed "working within the brigade framework and operating alongside a full range of capabilities was a useful education." Once deployed, the unit quickly adapted to local circumstances. Constant patrolling was the key, said Burnett. These patrols took many forms: reconnaissance, fighting, standing, presence, civil affairs, security, and engagement.[27] Burnet's approach fitted in well with that of the Australians. Confident of the commonalities, mutual understanding, trust, and respect, Burnett's immediate commander in the field, the Australian Brigadier Mark Evans, was confident in assigning a clear but broad mission for Burnett and his forces to pursue.

REFLECTIONS

This chapter has demonstrated that effective exercise of mission command during the East Timor crisis can be attributed to a number of factors. These include effective oversight, robust and clear command and control arrangements, strong and compelling leadership, and high levels of tactical proficiency combined with opportunities to build trust and mutual respect between contingents and force components.

For Australia, the East Timor mission also occurred when efforts to improve joint training and education were starting to bear fruit. The Australian Defence Force Academy (ADFA) had opened in Canberra in 1986 and its graduates were reaching the middle ranks of the respective services' officer corps (i.e., they were majors, lieutenant commanders, and squadron leaders). ADFA offered degrees in the liberal arts and sciences. It likewise also offered an opportunity for officer cadets from the three armed services to become friends more so than inter-service rivals. Profound benefits accrued in terms of improved teamwork and effectiveness even at the junior officer level. This was demonstrated in East Timor. What otherwise could have been a slow and tedious bureaucratic process for air support, for instance, was partly facilitated by familiarity and mutual friendship and respect.[28]

For Australia, INTERFET provided the necessary invigorating experience to generate an improved learning culture within the ADF. Yet it is important not to forget that INTERFET was not an opposed operation. Australia gained priceless experience in deploying and supporting operations; the situation conceivably could have turned out far worse. The prospect of fighting breaking out between INTERFET force elements and their opponents was a real one, and the fact that a general outbreak of further violence did not occur upon the arrival of the force was not simply due to good fortune. The stage had been set by effective international negotiations immediately prior to the deployment, but the prospects of success were bolstered by a clear understanding of the commander's intent and the political imperatives at work at all levels of the chain of command.

It is important to remember that the intervention presented a realistic prospect of armed confrontation with Australia's giant neighbor Indonesia—one that, given the vagaries and the "fog" of unscripted interventions with heated emotions and competing vested interests, pundits could not rule out possibly leading to war. The fact that such an outcome was avoided points to the determination demonstrated by both countries to exercise restraint, conscious of the long-term potentially catastrophic ramifications of doing otherwise. The fact that the outcome was so benign is in itself extraordinary. The crisis warrants close study, particularly in light of the dearth of other similar military interventions with such a positive ending.

As it happens, events that followed the onset of the so-called Global War on Terror in 2001 meant that the INTERFET experience faded in the public consciousness. On one level that is understandable, particularly given the virtual absence of INTERFET casualties compared with conflicts like the Vietnam War as well as hostilities in the Middle East. Subsequent events in Indonesia also contributed to the fading from public consciousness, to include the Bali bombings in 2002 and 2005, the Jakarta hotel bombings in 2003 and 2009, the Australian embassy bombing in Jakarta in 2004, and the staggering death toll from natural disasters such as the Indian Ocean tsunami in 2004.

These other events have tended to shade the intervention in East Timor, where the deployment easily could have gone horribly wrong. One trigger pulled by a nervous soldier at a checkpoint, or one irresponsible act could have escalated into a bloody clash from which it would have been difficult to extricate or back down. A less disciplined force, or a more trigger-happy one without a clear understanding of the commander's objectives, could have prompted an escalation of hostilities. Indeed, there was considerable concern about plausibly deniable acts of violence that could have derailed the intervention—particularly in the critical first few days from September 20, 1999 onwards. That this did not happen is a testament to the sensible, restrained, but firm hand exercised by Major General Peter Cosgrove and his multinational force as well as the recognition by Major General Kiki Syahnakri, his Indonesian counterpart, that this transfer of authority had to be made to work.

ENDNOTES

1   Map courtesy Australian National University College of Asia and the Pacific CartoGIS, map number 11-164/2JS.

2   There are numerous works outlining the history of events leading up to the East Timor crisis of 1999. The latest is John Blaxland (ed.), *East Timor Intervention: A Retrospective on INTERFET*, Melbourne University Press, Melbourne, 2015.

3   There are a range of books covering Australia's military involvement in the Vietnam War including an official history series. The latest, written by the official historian Peter Edwards, is *Australia and the Vietnam War*, New South, Sydney, 2014.

4   See Bob Breen, *Struggling for Self-reliance: Four Case Studies of Australian Regional Force Projection in the Late 1980s and 1990s*, SDSC, Canberra Papers on Strategy and Defence, No. 171, ANU E Press, Canberra, 2008.

5   D. Connery, "Lessons for Australia's National Security Policy and Policymaking from INTERFET," in Blaxland (ed.), *East Timor Intervention*, 95-96.

6   Ibid.

7   See Michael G. Smith, 'INTERFET and the United Nations' in Blaxland (ed.) *East Timor Intervention*, 13-25.

8   Photograph courtesy of Australian Defence Forces, image number ADF_V9925621.

9   See Mark Evans, "A Tactical Commanders Perspective" in Blaxland (ed.), *East Timor Intervention*, 115-126.

10   David Horner, *Making the Australian Defence Force: the Centenary History of Defence: Volume IV*, Melbourne, Oxford University Press, 2001, 157.

11   See John Blaxland, *The Australian Army from Whitlam to Howard*, Melbourne, Cambridge University Press, 2014.

12   This is the theme argued in John Blaxland, *Information Era Manoeuvre: The Australian-led Mission to East Timor*, Canberra, Land Warfare Studies Centre, Working Paper No. 118, June 2002.

13   C. Barrie, '*Creating an Australian-Led Multinational Coalition*', in John Blaxland (ed.), *East Timor Intervention: A Retrospective on INTERFET*, Melbourne University Press, 2015, 87.

14   P.J. Cosgrove, "Commanding INTERFET" in Blaxland (ed.), *East Timor Intervention*, 105.

15   Ibid, 108.

16   See Blaxland, *Information Era Manoeuvre*, 19-20.

17   P.J. Cosgrove, 'Commanding INTERFET' in Blaxland (ed.), *East Timor Intervention*, p. 111.

18   Ibid.

19   Ibid, 112.

20   Ibid.

21   Ibid.

22   Ibid.

23 David Horner, *SAS: Phantoms of War*, Sydney, Allen & Unwin, 488.

24 McMahon, discussions with author, cited in Blaxland, *The Australian Army from Whitlam to Howard*, 148.

25 Bob Breen, *Mission Accomplished: East Timor*, Allen & Unwin, Sydney, 2000, 57.

26 See Surasit Thanadtang, 'The significance of Thailand's contribution', in Blaxland (ed.), *East Timor Intervention*, 227-232.

27 Blaxland, *The Australian Army from Whitlam to Howard*, 157.

28 The author was in the last Duntroon class before ADFA. Colleagues from even one year below (with one year at ADFA with the other services) could effortlessly resolve inter-service problems (which appeared insurmountable to me) due to relationships established at ADFA. Such benefits are rarely considered by critiques of ADFA but make the investment invaluable.

8

THE SOLOMON ISLANDS INTERVENTION: THE REGIONAL ASSISTANCE MISSION 2003

Lieutenant General John J. Frewen, DSC, AM, Australian Army

BACKGROUND

In the years leading up to the 9/11 attacks in 2001, Australia was loath to intervene in the affairs of Pacific islands nations — even at their request — to avoid accusations of neo-colonialism. Despite repeated requests from Solomon Islands prime ministers, Australia had steadfastly avoided being drawn into the affairs of this small but geographically and tribally diverse nation prior to the 2001 assaults on New York and Washington, D.C. The exception was an unarmed International Peace Monitoring Team (IPMT) from November 2000 until June 2003. However, this intervention proved futile as it could not compel compliance by warring factions given its limited manpower, authority, and capability to coerce. Tribal militants simply hid their weapons and waited out a force restricted by its charter to no more than monitoring and reporting.

The events of September 11 changed this. Australia could no longer sit by passively while transnational criminals — and potentially terrorists — took hold in this neighboring country.[1] There was thus a dramatic shift in Australian policy mid-2003 . Australia responded positively to Prime Minister Alan Kemekeza's impassioned plea to help in resolving the downward-spiraling security situation in the Solomon Islands. Over the following few weeks, the force that would become the Regional Assistance Mission to Solomon Islands (RAMSI) was drawn together with regional and United Nations blessings and legislative backing from the Solomon Islands' Parliament. An Australian-led multiagency and multinational armed intervention force comprised of civilian, police, and military elements deployed by air and sea on July 24, 2003.

The force rapidly restored law and order, successfully bringing the troubled nation back from the brink of failure. RAMSI was ground-breaking in that the interagency elements—police, military, diplomatic, and aid among them—were fully integrated prior to its deployment under civilian leadership. The success of RAMSI was also a story of mission command, by both design and by necessity, within the diverse military force. As the commander of the coalition contingent, formal training and personal preference had made me an advocate of the style of directive control that mission command requires. As I was to learn, the short preparation time for the force and broad span of command made this style of command more necessary—rather than less—even if it required greater trust and risk on my behalf.

The Solomon Islands is an archipelagic nation with a population of less than 600,000, its villages are scattered across a chain of nearly a thousand medium-to-small islands. The climate and terrain is uncompromising. Deep jungle-covered mountains plunge suddenly into tumultuous seas and there are few open areas. Conditions are hot and tropical. Malaria remains rife. The capital of Honiara on Guadalcanal Island is the only city of note. Its legacy reaches back to the legendary United States Marine Corps landings on nearby Red Beach, where US marines fought the epic and bloody struggle against entrenched Japanese forces for control of Henderson Field, a rare piece of flat land little more than a mile distant from where the Americans came ashore. Over time, the airfield became a hub for the allies advance northward as the war progressed and the local people were increasingly drawn to the affluent and amiable US presence there.

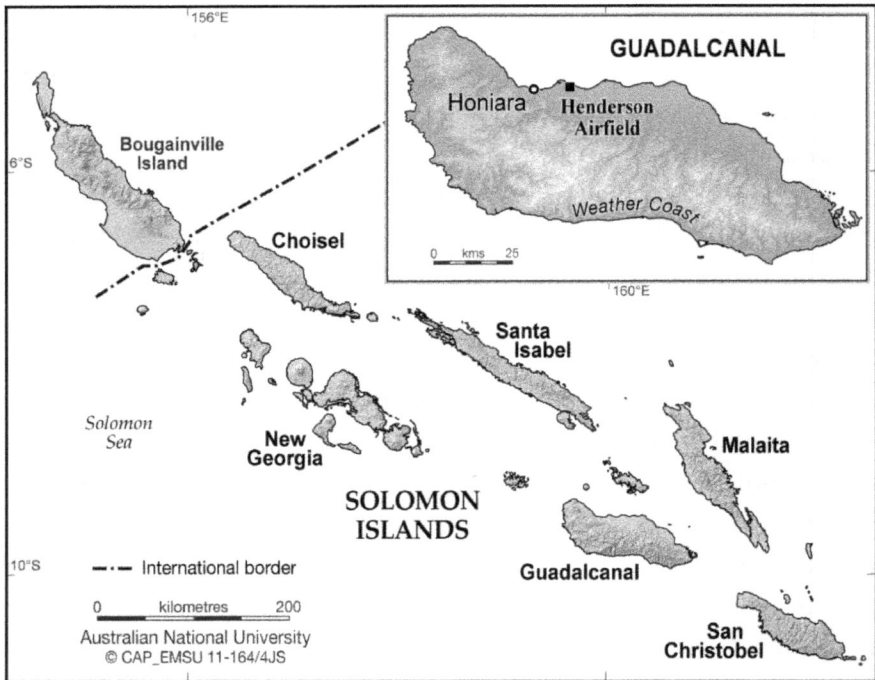

Map 8-1: Solomon Islands[2]

Honiara became the center of national enterprise after the war. This increasingly attracted the natives of Guadalcanal—Gwales (goo-wa-lees)—and later Mailaitans (May-lay-chi-ans), the latter form the colony's other main but smaller tribal group from the island of Malaita. The Malaitans' enterprising nature and increasing numbers saw them come to dominate business and positions of authority. Tensions escalated between the two protagonist groups after the colony gained independence in 1978 and continued throughout the 1990s. Major militia organizations; The Guadalcanal

Liberation Front (GLF) and the Malaitan Eagle Force (MEF) formed in response and entered into conflict. Resultantly, more than 20,000 Malaitans were driven back to their home island and violence persisted through the time of the "toothless" IPMT's tenure up to the time of RAMSI's July 2003 arrival.

Australia took advantage of lessons learned from recent peace support operations in Bougainville, East Timor, and elsewhere in forming this multi-faceted task force. These earlier contingencies helped mission leaders appreciate the importance of fully integrating multi-agency capabilities prior to deployment. A civilian special coordinator, career diplomat Nick Warner, was the overall lead for the mission. Assistant Commissioner of the Australian Federal Police, Ben McDevitt, led the combined participating police forces. A public servant from the Australian Agency for International Development (AUSAID), Margaret Thomas, headed the relief and development group. I commanded the almost 1,800 strong joint military force from five contributing nations (Australia, New Zealand, Fiji, Tonga, and Papua New Guinea). Unusually, the larger military force was in support of the police (who numbered around 200 for the first few months of the mission) and other organizations from the ten countries that together comprised RAMSI (which totaled a little over two thousand). These unique aspects of the RAMSI mission were central to our prospects for achieving an enduring solution and mission command would be necessary for achievement of our mission.

RAMSI's all-agency command structure defied the orthodoxy of initially deploying military stabilization forces and subsequently transitioning authority to police and civilian agencies. Security requirements, establishment of law and order, aid distribution, and bureaucratic assistance all progressed steadily from the earliest days of the mission. This facilitated earlier support to the Solomon Islands and a swifter return to normalcy. The integrated approach underwrote the long-term success of a mission that was to last over ten years and saw an ever-increasing focus on non-military assistance within weeks of the initial intervention. In fact, the military force was dramatically reduced after only four months (my command tenure). In over little more than 100 days, RAMSI restored security, arrested key warlords, accepted the surrender of the two key rival militias, collected and destroyed over 2,500 weapons, and seized hundreds of thousands of rounds of ammunition across an extended archipelago. Without firing a single shot, RAMSI restored the confidence of a struggling nation and secured the local population's overwhelming support. Australia's approach to mission command underpinned this success at every level.

Figure 8-1: From left— LTCOL John Frewen (Australian Army), Nick Warner (Department of Foreign Affairs and Trade), Ben McDevitt (Australian Federal Police)[3]

PRE-DEPLOYMENT DECISIONS

RAMSI typified mission command at the strategic, operational, and tactical levels. Nationally, then Australian Prime Minister John Howard empowered the heads of the four arms of the mission through individual letters of appointment. Warner was directed to coordinate the efforts of the agencies toward the common goals of delivering peace, security, and prosperity. In my case as military lead, the letter stated I was to retain full control of all military forces but should remain responsive to the requirements of both the special coordinator and head of the participating police forces. Strategically, our instructions were unambiguous and concise. They provided clear freedom of action within delegated lines of responsibility. I also personally received verbal guidance from the Chief of the Australian Defence Forces (ADF), General Peter Cosgrove (recently Governor General of Australia). He laid out the aim of the mission from the military perspective, the tone he expected in terms of how military personnel should conduct themselves, constraints regarding the way operations should be conducted, and timeframes for the anticipated duration of Australian Defence Force support. He added that that this mission would "set the face of ADF operations in the Pacific for decades to come" and that I had "better not f*%# it up."[4] His strategic intent was crystal clear and required no further elaboration.

Within days, Australia's operational headquarters, then known as HQ Australian Theatre (HQ AST), provided additional planning guidance—including some parameters such as freedoms and constraints—and an intended force structure. This HQ was now my direct command HQ and I was responsive to it rather than my parent brigade from this time on.[5] It was reassuring to note that there was already clear intent to deploy a capable combat force with a potent mix of joint enablers. Central to the mix were a light infantry battalion supported by rotary and fixed-wing aircraft, a range of naval vessels including a landing platform amphibious (LPA), and a comprehensive suite of intelligence assets among others. (See Figure 8-2.) Despite the breadth and depth of resources, I was somewhat concerned in having a force structure designated before I had a plan. This was done in part because the operational headquarters was aware of the potential mission for some weeks before I was advised. We were now less than three weeks from insertion.

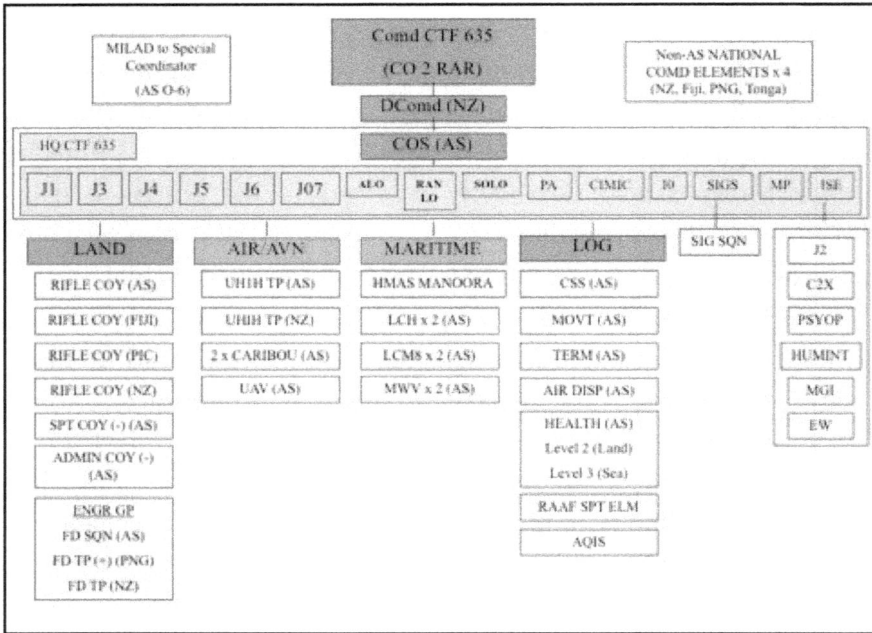

Figure 8-2: Order of battle for first RAMSI rotation of Combined Task Force 635, July 24—November 18, 2003[1]

1   Acronyms: 2RAR = 2nd Battalion, Royal Australian Regiment; ALO = air liaison officer; AQIS = Australian Quarantine Inspection Service; AS = Australia; C2X = coalition intelligence staff section, counterintelligence and human intelligence; Caribou = DHC4 transport aircraft; CIMIC = civil-military coordination; CSS = combat service support; EW = electronic warfare; HQ = headquarters; HUMINT = human intelligence; IO = information operations; ISE = intelligence support element; J1= personnel staff section, joint headquarters; J2 = intelligence staff section, joint headquarters; J3 = training and operations staff section, joint headquarters; J4 = logistics staff section, joint headquarters; J5 = plans staff section, joint headquarters; J6 = communications staff section, joint headquarters; J7 = engineering staff section, joint headquarters; HMAS Manoora was a modified LST=landing ship tank. LCH = landing craft–heavy; LCM8 = landing craft–medium, model 8; MGI = military geographic information; MILAD = military advisor; MP = military police; MWV = minor war vessel; NZ = New Zealand; PA = public affairs; PIC = Pacific Islands Contingent; PSYOP = psychological operations; RAAF ELM = Royal Australian Air Force element; RAN LO = Royal Australian Navy liaison officer; SIGS = signals; SOLO = special operations

I enquired as to what flexibility I had in amending the allocations anticipating that the details of the mission would become clearer over the next several days as our planning progressed. The response was that the allocations were pretty well fixed but I could tinker in the margins less a few capabilities that were not negotiable. One I could not "tinker" with was the composition of the thirty-person strong civil-military cooperation element based on General Cosgrove's experience of having lacked such a group in East Timor. As the operation progressed, it transpired that the preordained force structure was more-good-than-bad and, on reflection, its design by the operational headquarters was appropriate. It resulted in a better range of joint assets than I might have requested, and thanks to the headquarters preliminary planning, the units identified were ready to deploy.[6] I was nevertheless concerned with having been allocated only three light infantry companies rather than the four organic to Australian light infantry battalions at the time. I argued this would be insufficient for sustained security operations in a dispersed theatre but was overruled.[7]

At this time I was a junior lieutenant colonel in my first year as commanding officer (CO) of the 2nd Battalion, The Royal Australian Regiment (2 RAR), the core of Australia's "on line" ready battalion group (RBG). The RBG was in-turn part of the Ready Deployment Force (RDF) based on the army's 3rd Brigade. Previous ready battalion groups had deployed to Somalia and Rwanda, and had led the intervention into East Timor as part of a RDF brigade group in 1999. 2 RAR was well drilled and had only a few months earlier rapidly deployed a company group to Iraq to establish the first of what became a longstanding security detachment to Australia's Baghdad embassy.[8] Officers within the RBG and RDF were carefully selected and were typically well known to each other. Due to exercising and training, and a propensity for a majority of

liaison officer; TERM = terminal detachment; UAV = unmanned aerial vehicle; UH1H = model of helicopter.

my officers to spend their regimental careers within the RDF, the strengths and weaknesses of the officers within the RBG group were particularly well known to me. With few exceptions, I had high confidence in them and there were strong foundations for mission command.

Although the Solomon Islands mission was well suited to the RBG, the span of command, command structure, and the interagency nature of the mission was new ground. The range of joint assets could not be effectively managed by a light infantry battalion headquarters and was clearly beyond that normally accorded to such a junior officer. This posed a number of immediate challenges that would require innovative thinking which arguably could only have taken place in a mission command environment. However, my ability to exercise mission command would also be tested by the integration of foreign troops whose leaders I did not know (let alone have a feel for their understanding of mission command). With the time available I had no choice but to be explicit in my instructions and to trust them until given reason to do otherwise.[9]

The first pressing task was to modify the ready battle group and my headquarters to become capable of coordinating air, land, and sea operations across a diverse and extensive area of operations. Fortunately, I was familiar with combined joint task force (CJTF) operations after exercising with the US I Corps and III Marine Expeditionary Force (III MEF) headquarters during my exchange posting as a planning major in Headquarters, US Army Pacific (USARPAC) some years earlier.[10] I drew on this experience to structure my force and headquarters appropriately prior to deployment. I did so along environmental component lines rather than as is normal for a battle group and its staff.

I formed air, maritime and logistic components that were respectively commanded by: Major Peter Steele, an aviation staff officer in my HQ; the Navy CO, Commander Martin Brooker, aboard the Her Majesty's Australian Ship (HMAS) *Manoora;* and

Major Donna Boulton who commanded the logistic elements from the airfield. I retained command of the land maneuver components, including the rifle companies and the engineer group commanded by Major Jason Hedges. Finally, I used my supporting artillery battery commander, Major Charles Weller, an individual well known to me, as a CJTF chief of staff (COS) to manage the HQ staff (who was otherwise without portfolio as no artillery pieces were deployed to the Solomon Islands).[11] I was given freedom to arrange my force as I saw fit and formed a HQ with what I believed was the greatest chance of successfully coordinating our mission despite our unorthodox structures and limited preparation time.

I was also "hatted" as an agency lead within the context of our mission. Once deployed, there was an almost daily requirement for me to attend interagency coordination meetings, government or public briefing sessions, media interviews, and so on. These obligations soon consumed half of my days, drawing me away from spending as much time with our soldiers as I would have liked. The span of command compounded the challenge. There were eighteen subordinate units from four countries (New Zealand, Fiji, Tonga and Papua New Guinea) under my direct command in addition to those from Australia. I initially tried to manage them as I would the six sub-units of an infantry battalion, an approach that quickly proved unwieldy and ineffective. I therefore relied on my trusted COS and headquarters staff to coordinate daily operations and oversee sub-elements in accordance with my intent. The tempo of operations and dynamic nature of our mission reinforced the requirement for exercising a mission command approach down to the lowest tactical elements which were often geographically remote and isolated.

In particular, I relied heavily on my COS, Charles Weller, and my battalion S-3 (operations officer), Major Dave Smith (who was now designated as the J-3 given the joint character of the command), to coordinate the day to operations within the command post.[12] Both officers were highly competent majors and well known to me through demanding training. I had total confidence in their expertise and reliability and could fully trust them to meet my intent. Similarly, the HMAS *Manoora*'s commander and I had established a rapport during an amphibious exercise conducted as part of our routine build-up training. I therefore had confidence in his ability to run the command's maritime component—conducting interdiction operations and the like—in parallel with our other joint tasks. The eighteen land-based sub-units were a different matter. Some commanders I knew well; others I did not know at all. As some were from other countries, I was also less familiar with their modes of operation than was the case with Australian Army organizations. I knew I initially had to pay more attention to those with whom I was less familiar until we established a mutual understanding of each others' capabilities and they demonstrated the command qualities fundamental to gaining my trust. The compressed timeframe of our deployment combined with the intense tempo of early operations meant I could not commit as much time as I would have liked to building these relationships. Knowing that both international and Australian military authorities sought to deploy only the best of their leaders and units, I was comfortable in assuming these subordinates would deliver. Nonetheless, those less familiar received my closer attention—and that of my staff—and it was essential we moved quickly to rectify any problems that arose.[13]

A final friction point in some circles was the CDF's decision to appoint me as the combined-joint task force (CJTF) commander despite my role as the RBG commanding officer. Many within the ADF felt that, at a minimum, an ad hoc colonel (O6)-level headquarters should have been created for the mission. Others believed that my parent brigade headquarters should have lead the mission. This created a propensity for intense external scrutiny of my tactical decisions from Australia. Critiques by distant naysayers unhelpfully amplified what were already significant command pressures as the mission unfolded. Anonymous criticisms continued until the outcome of the mission was no longer in doubt—and although irritating—they pleasingly did not manage to undermine the strong support from my direct chain of command.

I have never determined the full range of reasons why the CDF appointed such a junior officer commander of RAMSI's military contingent. I believe contributing factors included the defense force being unable to deploy a brigade-level or purpose-built headquarters while concurrently being committed to significant support of deployments in East Timor and Iraq in addition to a potentially expanding commitment in Afghanistan. It may also be that General Cosgrove was concerned that a more senior officer may have been less inclined to defer to the numerically smaller police contingent. It was certainly my good fortune to be the RBG commander at the time of the Solomon Islands deployment. However, it was also fortuitous that I had worked for the CDF as a captain running the infantry platoon commanders first appointment course when he was the colonel commandant of the Australian Infantry Centre. This familiarity with my past performance most likely also contributed both to his confidence in assigning me command and the freedom of action granted me in his informal commander's guidance. His intent was formalized, and then strictly accorded to, through the series of back-briefs and final approval of my plan by HQ AST. Every level from strategic to tactical was clear about what we were to achieve and the nature in which we were to do so.

Despite years of reflection, I am uncertain whether a different command and control arrangement would have improved the tactical achievements of RAMSI's military contingent. As I was to learn, you cannot argue with success. Perhaps a less favorable result would have led to greater criticism of the mission's command structure...and its commander. Regardless, the challenge was great, its demands difficult, and acceptance of significant risk not an unusual necessity. I firmly believe that our commanders, other junior leaders, and me greatly benefitted from the increased levels of responsibility entrusted to us during those first months of RAMSI and that those experiences continued to pay dividends for Australia's military forces and those of our partner nations in the years following. Mission command properly executed not only makes for effective operations today; it prepares those entrusted for missions yet to come.

DEPLOYMENT PREPARATIONS

The military component of RAMSI was forewarned, formed, and deployed over a period of less than three weeks. It was paltry time to assemble a coalition and plan, train, and assimilate with other agencies and nations. The briefness of the period also left little opportunity for reconnaissance or completion of thorough logistical arrangements. The sheer size of Australia and dispersed locations of various headquarters and supporting elements meant a considerable part of those three weeks was spent travelling by myself and key CJTF staff. For example, I had to travel from Cairns in northeastern Australia (where I was on leave) south to Sydney for my initial orders, return north to Townsville to issue planning guidance, visit the Solomon Islands for a reconnaissance, return to Sydney to back-brief my plan and obtain the approval of the Commander, Australian Theatre (COMAST), return to Townsville to issue orders, then return south to Canberra for planning with the Australian Federal Police (AFP) and other agencies in addition attending HQ the Strategic Command Group[14] before once again returning to Townsville. This involved more than 20,000 kilometers of air travel in combination with myriad other requirements. Fortunately, my trusted staff assumed much of the burden and support from throughout the ADF was topnotch. The effect was nonetheless exhausting. I was longing for the few hours of sleep on the C-130 en route to Henderson Field by the time D-Day arrived.

I was in the Solomon Islands for seventy-two hours as part of the Solomon Islands reconnaissance mentioned above. Frustratingly, I had to argue to travel to Guadalcanal with several of my staff. The original intent was for HQ AST to conduct the visit without representation from 2 RAR. This was wrong-headed but eventually I was included in the party with my engineer squadron commander and anticipated logistics component commander. A communications representative would also have been beneficial in hindsight, but we were lucky to get the positions we did.[15] The visit was highly productive and proved critical to my understanding of the operational environment and to shaping my insertion plan. My presence also helped alleviate concerns within elements of the Solomon Islands establishment, such as the Solomon Islands police commissioner, who had reservations about military operations, concerned that we might deploy too aggressively and create a backlash by "kicking-in doors around town."

Among my other concerns were two CDF stipulations in particular. First, unlike in East Timor, there were to be no national areas of operations. Second, I was to conduct the operation consistent with an ability to draw down the military force after ninety days. A New Zealand liaison officer fortunately joined our reconnaissance during the time in Solomon Islands and began working to establish an independent area of operations. The well-timed opportunity for face-to-face discussions allowed me to pass the message that the mission would be fully integrated down to the lowest tactical levels, thereby heading off a disagreement that would otherwise have had to be resolved on D-Day. My brief time in Honiara also allowed me to detail my tactical dispositions, something that would have been impossible from Australia. General Cosgrove's clear intent and relatively few but quite specific constraints provided me the freedom to dictate the full range of tactical details for our joint and multinational force (pending final approval by HQ AST, of course). So too, requirements to backbrief my plan at crucial stages reflected an appropriate balance between decentralization and monitoring a fairly inexperienced commander readying to deploy on an operation with potentially dramatic and long-lasting regional implications. This was effective mission command from higher commanders.

A decision not to allow me representatives on the pre-deployment reconnaissance would have been an error of judgment but not a mission command failing. Yet the opportunity was crucial to my exercise of mission command. I returned to Townsville having made my intent clear to our New Zealand partners and far more comfortable that I could better define my own intent regarding the positioning of and tasks to be undertaken by RAMSI's armed forces component. My enhanced knowledge also allowed me to set in motion negotiations for leasing facilities in readiness for our arrival.

I was, however, still uncertain as to how we might be able to neutralize the militant warlord Harold Keke. Keke hung like a dark specter over RAMSI's success. He was an almost mystical figure ensconced on the remote Weathercoast of southern Guadalcanal. Keke had formed and continued to lead the GLF, a force instrumental in driving Malaitans back to their home island. He had since become a despotic and murderous figure, waging an ongoing war with the rival Istabu Freedom Movement (IFM) along the Weathercoast while ruling his own people with an iron fist. Little was known about him and his militants. Virtually no one dispatched to negotiate with him had returned alive. How we would capture Keke without being drawn into a protracted fight in his home patch of steep and densely vegetated jungle was a dominant concern, one that remained unresolved in the aftermath of my reconnaissance.

With a personal sense of the situation on the ground, I was able to alter some views held by senior officers in Australia regarding how the mission should unfold. I argued against the inclusion of armored vehicles for two reasons. First, I had found during my experience with the United Nations in 1994 Rwanda that although armored personnel carriers could be reassuring in some situations, on balance they alienated local forces with the result that greater numbers of opponents carried heavier weapons such as RPGs. Second, I believed that images of armored vehicles on the streets of Honiara would reinforce a perception of the Solomon Islands as a failed state, delaying the return of tourists and investors. The tone of the mission had to be one that sought to facilitate stability's swift return. I was also assured by the Commander, AST that he would have light armored cavalry vehicles (LAVs) on the ground overnight if I felt the situation demanded them. The decision relieved pressure on our HMAS *Manoora* loading plan and air bridge priorities. Subsequent events fortunately vindicated the assessment.

While I was seeking approval from higher headquarters, the AFP and other agencies were conducting parallel planning at various distant locations. I had deployed liaison officers (LOs) wherever I felt it appropriate to ensure we were aligned with these vital elements but at this stage there was too much uncertainty to rely on this means of coordination alone. We were still discovering significant misalignment of capabilities and misunderstanding between RAMSI participants though deployment was but days away. Our saving grace was a whole of government rock drill. Myself and Colonel Paul Symon, the Special Coordinator's Military Adviser,[16] pressed for this despite some resistance, and the event proved critically important. The synchronization rehearsal included all participating and supporting agency leaders along with those from academia who assisted us in better understanding the operating environment. This activity was, to the best of my knowledge, a form of inter-agency integration previous operations had not undertaken. It was to prove one of the most significant activities we conducted as a multi-jurisdictional and multi-disciplinary group.

This interagency rehearsal helped all participants gain a common understanding of the plan. It also was invaluable in helping us to synchronize our efforts during the first days of the mission. It revealed that we risked losing the Solomon Islands' population's perception of momentum. We recognized it would take a month or more to gather sufficient evidence for prosecutions even after restoration of security. This led to development of RAMSI's weapons amnesty program involving a nationwide tour to announce our arrival and the requirements with respect to turn-in of illegal weapons. We also developed a number of open day-style events to build rapport with the local community, showcase our capabilities, and leave no doubt we could—if necessary—hunt out weapons and overwhelm any who tried to oppose us. Without the amnesty program we risked floundering and squandering early good will.

The short pre-deployment timeframe did allow for the usual range of individual preparations such as zeroing of weapons, inoculations, and cultural briefings but did not leave adequate occasion for collective training. In fact, while military forces from the three Pacific Islands nations concentrated in Townsville and deployed with us, most New Zealand elements joined us directly in the Solomons. We thereby assumed significant risk given the possibility of facing opposition during the first hours of the insertion.

The combined joint task force headquarters (CJTF HQ) built on my battalion headquarters faced numerous challenges despite the beneficial preparations. We were fortunate in having a command post exercise (CPX) provided by Australia's Combat Training Centre (CTC). Though only a paltry six hours could be afforded during the final precious days prior to deployment, the CPX had a profound effect in facilitating important last minute amendments to plans. One such adaptation was the decision to collocate the police and military operations cells within the headquarters. The AFP's initial desire was to have a separate police call and monitoring center; the CPX demonstrated this might cause confusion and delay provision of military support during a crisis event.

Not surprising given that ours was a battalion on a very steep learning curve, the CJTF headquarters was deemed to be "underdone" by the CTC evaluators. At that stage we-were-what-we-were; the time for even limited training was over. One saving grace was that the core CJTF military staff knew each other well from years of previous service and training. This was less the case with the foreign forces but there were some links that helped. For example, I had served on exchange in New Zealand for two years training their new officers. This meant I understood their training priorities and standards well and either knew some of the kiwi RAMSI officers personally or had close mutual associates that reinforced a sense of trust. I had also trained PNG officers while running our infantry platoon commanders course. This meant I understood their capabilities and, fortunately, had a good reputation that had preceded me with the PNG troops. Small and serendipitous links at times like this can help the implementation on mission command across national groups.

Subsequently, we were all confident that we understood our mission and the freedom of action granted us. However, it would be weeks after the insertion before the military and police fully understood each other to the extent necessary and months before similar bonds of trust to those across the military developed between individuals from different agencies. In a practical sense this meant that while I could usually trust the military representatives of my command to act in accordance with my intent, we had to pay particular care to ensure inter-agency efforts remained well-aligned. This was a dimension beyond our understanding of mission command as described in Australian doctrine, as aspect that cannot be assumed away when agencies come together.

DEPLOYMENT

Shortly after dawn, at 0700 hours on July 24, 2003, waves of C130 aircraft commenced landing at Honiara International Airport. Simultaneously, the Australian amphibious landing ship HMAS *Manoora* crested the horizon, announcing her presence to the people of Honiara. Shortly thereafter, army LCM8 flat-bottomed ship-to-shore landing-craft began ferrying troops, police vehicles, and equipment from HMAS *Manoora*, disgorging them at nearby Red Beach. Troops exited the aircraft without body armor, weapons slung, waving to the thousands of ecstatic locals pressed against the tall wire fences surrounding the airfield. In the week prior, media reports had indicated that RAMSI soldiers, unlike the unarmed IPMT, had "shoot to kill" rules of engagement (ROE). This is not language the ADF typically used to describe ROE, but it had conveyed a clear message of deterrence prior to our arrival. We had named our undertaking *Operation Helpem Fren,* local pidgin for helping a friend. This sought to unambiguously reassure those in the Solomon Islands regarding our intent. By arriving with a friendly but alert demeanor, we sought to impress that we had come to help but were ready to deal with any hostile intent militias or criminals who might demonstrate.

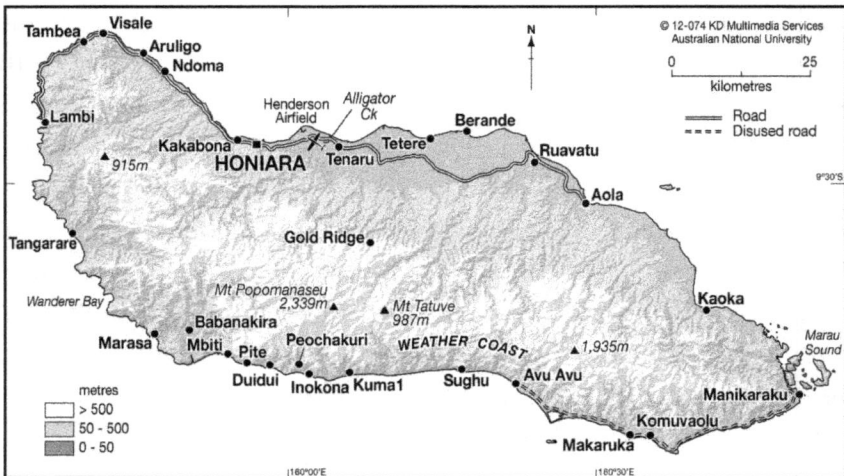

Map 8-2: Guadalcanal[17]

Fortunately, no nefarious elements decided to resist our arrival. We had secured the abandoned resort at Red Beach by the end of the first day. (Red Beach was the scene of the heroic and costly landings by USMC forces some sixty years earlier.) Honiara's airport was built on what had previously been Henderson Field, the ultimate prize of the WWII struggle for Guadalcanal. Australian and Fijian rifle companies were in security positions as dusk arrived on D-Day. Our nascent CJTF headquarters had established control. Most importantly, RAMSI law enforcement officers were in the Honiara police station and had commenced patrolling with Solomon Islands police within hours of arrival. All this had been achieved without major incident or violence.

Figure 8-3: Australian Army soldiers patrol along a Guadalcanal beach, July 28, 2003[18]

OPERATIONS

Activity continued at a frenetic pace in the months that followed deployment. RAMSI leaders succeeded in maintaining the mission's momentum and, with a few hiccups, remained overwhelmingly popular with the general population. The initial priorities for the military contingent included:

- Maintaining security for RAMSI personnel and facilities
- Facilitating leadership visits to all corners of the Solomon Islands to spread word of the mission and the requirements regarding the weapons amnesty program
- Establishing and building—and where necessary protecting—police outposts at strategic locations throughout the islands
- Gathering and destroying weapons and ammunition in support of the amnesty program.

Members of the local population turned-out in droves, particularly women who, as ever, carried the burden of social unrest and responded enthusiastically to the destruction of weapons.

Figure 8-4: Solomon Islanders lining Honiara street during August 25, 2003 visit by Australian Prime Minister John Howard[19]

Having assembled quickly and not having previously worked closely together, there were misunderstandings between agencies during this opening phase. However, those at every level worked hard to overcome the obstacles. Not being the lead agency was new terrain for most of us in the military. I constantly reinforced our supporting role with my troops. Conversely, I needed to reinforce with supported agencies a need to articulate support requirements in terms of "effects not assets" (e.g., request movement support rather than a helicopter specifically). This was a difficult road. A mismatch in interagency expectations regarding each others' capabilities combined with the absence of a memorandum of understanding between the ADF and the AFP caused tensions that could have been avoided with greater mutual understanding. These were overcome with time but were noted for future improvement.[20] Other agencies did not have mission command doctrine similar to ours but acted within similar constraints nevertheless. The AFP lead, Ben McDevitt, had previously served as a soldier, knew his Australian police officers personally, and trusted them to act within his guidance. He quickly extended this trust to foreign police, albeit with appropriate controls depending on individuals' demonstrated expertise and established reliability.[21] Special coordinator Nick Warner did not have previous military service but maintained a collegial approach to interagency coordination and delegated well. That Nick and Ben had previously worked together closely in the counterterrorism domain benefitted the mission. However, it did mean that initially they were initially less trusting of those in RAMSI's military component. We were not known quantities to them. Trust could only be developed over time.

It is remarkable that things proceeded as well as they did given the need for speed in deploying, difficult operating conditions, and extent of interagency cooperation. It was a clear overarching unity of purpose that kept commanders and other organizations' representatives from each agency on a common track. I had confidence that I understood the parameters for our military operations. Only rarely did I need to seek guidance from my higher headquarters or find we collectively as RAMSI's leadership could not reach consensus without referring back to Australia.[22] Collegiality became a byword in the interagency domain. As when any plan meets reality, however, logistics soon worked its way to the fore.

Our logistical build-up continued, predominantly via the air-bridge, requiring some tweaking to balance police and joint force requirements as the mission progressed. The police's priorities had not been fully understood during pre-deployment planning. Reshuffling was therefore required to ensure timely delivery of police vehicles, equipment, and uniforms. I quickly learned much about joint military, interagency, and multinational logistics as we worked to bring order to what flowed into the country. Notably, our priorities were overturned early in the mission when unseasonal rains threatened to sink our encampment at the former beach resort. WWI style wooden duck-boards became my top priority to avoid all of us disappearing into deep mud. Whether in burning sun or monsoonal rains, the build-up continued and troops continued to live in make-shift tent lines for months as priority was given to the more permanent facilities police required to operate among the civilian populace. Some logistical decisions had broader implications and were based on lessons from previous ADF regional operations. For example, I decided that the vast majority of our rations and other consumables should come from Australia. This drew criticism from local representatives who had anticipated an economic boom thanks to our arrival. I made the decision to avoid unduly stressing the local economy or denying the local population staples. The policy also aimed to prevent creating false economies that would collapse once RAMSI drew down. Fortunately, no external pressure—political or otherwise—restricted our freedom of action in this regard.[23]

RAMSI leadership conducted visits to over fifty locations throughout the islands during the operation's initial weeks by vehicle, helicopter, boat, or a combination of all three. We found ourselves in urban settings, on beaches, and in deep jungle. Public radio occasionally broadcast the proceedings, to include question and answer sessions with local islanders. We conducted several military displays and demonstrations attended by thousands of curious citizens. The net result was that the entire country knew what RAMSI was within weeks of our arrival, who the faces behind its policies were, and that we had the military capabilities and supporting legislation to find and seize weapons and apply harsh penalties of up to $25,000 in fines or ten years imprisonment for noncompliance. Risk of punishment clearly weighed on people's minds, as did the opportunity for peace. Weapons were surrendered in droves. Employing a lesson learned during earlier operations in Bougainville (where rumors circulated that seized weapons had been reissued to various opposing forces), we committed to destroying weapons in the communities handing them in.[24] Word quickly spread that we were determined to end the rule of the gun.

HMAS *Manoora* was the jewel in our crown throughout our operations. This highly capable LPA provided not only very versatile ship-to-shore surface options. The ship also had two SK-5o Sea King helicopters that could penetrate weather our Australian and New Zealand UH-1H Iroquois helicopters could not.[25] This was a critical asset on a number of occasions, particularly when linked to our sole surgical facility aboard the HMAS *Manoora*. Our combat power and robust ROE were essential to success in RAMSI's early stages, but it was the LPA that really provided our ability to maneuver effectively during the period she remained in-theatre. Without *Manoora*, we could not have pulled off our greatest coup — the arrest of Harold Keke.

Figure 8-5: A local Solomon Islands police constable and RAMSI military and police representatives hold destroyed firearms after the first weapons destruction ceremony in Honiara. From left to right: Private Tuate Karikaritu (Republic of Fiji military forces), Lieutenant Colonel John Frewen, Police Constable Simon Tofuola (Royal Solomon Islands Police), Sapper Craig Reedman (3 Combat Engineer Regiment, Australia Army), Solomon Islands Police Commissioner William Morrell, Assistant Police Commissioner Mark Johnsen, and Sapper Stuart Pollock (2nd Royal New Zealand Engineer Regiment)[26]

HAROLD KEKE

The head of the RAMSI participating police, Ben McDevitt, had by way of clever police work and use of interlocutors made contact with Harold Keke in his jungle hideout to arrange a meeting. Harold's village, Mbiti, on the remote and inhospitable Weathercoast of southern Guadalcanal, was traditionally only accessible by four-day foot trek or perilous canoe voyage. Helicopters could land on the thin stretch of beach but were then at risk of being stranded by the frequently vile weather. Our plan to meet Keke required the HMAS *Manoora* to remain offshore (unable to anchor because of the currents) while the "Big Three" of RAMSI leadership were flown ashore (Nick Warner, Ben McDevitt, and myself). Troops

230

would remain at the ready both aboard the ship and in the air in the second Sea King should the meeting prove an ambush. As the security adviser, I could offer no more sophisticated advice to my two fellow leaders than "run into the ocean and wait for someone to fish you out" if all hell broke loose. It was not a strong plan, but I could see no better alternative with only a thin strip of beach backed by steep, dense jungle and no telling how many armed GRF members who were on their home turf.

We landed and were met by Harold's men. They advised us that Harold would not meet us that day (although we later learned that he had been watching from a concealed position). We flew back to the *Manoora*, consulted, then sent Colonel Paul Symon back ashore to issue an ultimatum that we would return a few days later and that Harold was to be present as he had previously agreed. Harold did appear during our second visit. He was a striking and strange character with a raspy, almost mesmerizing, voice. The meeting venue was a small chapel with rambling messages and odd diagrams scrawled on blackboards. Flanked by his senior lieutenants, whom he occasionally consulted but who remained thoroughly deferential to him, the scene reminded me of an odd mix of Conrad's *Heart of Darkness* and the film *Apocalypse Now.* The three of us had earlier rehearsed for the meeting and had a plan for the negotiations, but the setting was more unusual than we had anticipated. Notwithstanding, Nick led the negotiations with Ben highlighting the gravity and implications of the criminal behaviors that had transpired. My role was to be a reminder of the latent military might of RAMSI that could be brought to bear. By meeting's end Harold had agreed to have us return again to take him in to custody (along with his young wife, infant child, and his closest—and also murderous—lieutenant, Ronnie Kawa).

Why Harold agreed to surrender only three weeks after our arrival and without a fight has been the subject of much conjecture. The reasons were complex. Harold told Nick Warner (who at this point was known throughout the Solomon Islands and had had children named in his honor) that God had told him in a dream that Nick was a man he could trust. Harold was also adamant that the Solomon Islands government was responsible for all the country's "troubles" and he wanted to have his fair hearing in court (supported by reams of so called evidentiary documents they displayed for us). Additionally, by this time Weathercoast violence between the GLF, Harold's own village, and other groups including the IFM had escalated to such an extent that a circuit breaker was required. RAMSI offered this prospect as a trusted intermediary. The final reason was that the HMAS *Manoora*, although not a warship, loomed on the horizon as a very visible symbol of power seemingly akin to that of the USS *Missouri*. With helicopters circling above the *Manoora*, the deliberate impression was that our military could be used to hunt down Harold and his men if necessary. It was in this context that Harold Keke chose safety and justice over further conflict.

There was criticism from our coalition partners in Honiara that we had met with Keke and not taken the opportunity to seize him on the spot. This was the first (and only) time that the consensus of the coalition nations was challenged.[27] The concerns fortunately proved misplaced. I have never forgotten the moment of elation when Harold waved to the hundreds of villagers on Mbiti beach and stepped onto our Sea King helicopter for the short flight out to HMAS *Manoora*. We knew that everything would be easier from there on. No number of "critics" from afar could make a dent in what we had accomplished by neutralizing the most dangerous of the threats to Solomon Islands stability. I also knew that we would not have to wage a costly war against the well-armed native fighters of Guadalcanal's southern coast. What Harold did not know was that once aboard the ship he would be charged with the theft of a motorboat engine and detained until more substantive evidence could be assembled in pursuit of what were ultimately convictions for other crimes that included murder. The result was multiple lifetime sentences.[28]

The atmosphere was electric when we returned to Honiara. The news of the surrender rocketed around the Solomons and wider Pacific region. With Keke in prison, the key threat to other groups was removed and hope was restored. Harold and the GLF had been the final impediment to other groups surrendering their weapons. Within weeks the Guadalcanal Revolutionary Army (GRA), MEF, and IFM had declared hostilities at an end and surrendered their arms during ceremonial events. There was still much work to be done, but RAMSI had accomplished more than many expected, done so far more quickly than any had dared imagine, and achieved these ends without bloodshed.

CONCLUSION

The military component of Operation *Helpem Fren* wound down dramatically after four months, at which point I returned with the majority of 2 RAR to Australia. The force dropped from just under 2000 to around 800. A few weeks prior, HMAS *Manoora* had returned to Australia, her crew returning from a Middle East tour extended when she had been directed to support us in the Solomon Islands. RAMSI's military component would be down to less than 400 a few months later. Civilian contractors would provide capabilities such as helicopters. Though slightly longer than the anticipated ninety days originally envisioned by General Peter Cosgrove, the result was essentially in-line with his earliest guidance to me. The five nations' armed forces had been essential to deploying and sustaining the mission and to its greatest tactical successes. This could be attributed to both practical support and the application of important lessons drawn from other regional operations in preceding years.

The evolution of mission command as ADF doctrine was an unheralded but essential factor in RAMSI's success. General

Cosgrove trusted me, a junior officer, and the 2 RAR with a complex mission beyond the normal scope of our training and capabilities. He did this despite the strategic risk and the longer-term importance of the task. A substantial and potent mix of joint and coalition assets was allocated to mission accomplishment; I, as the tactical commander, was accorded freedom of action with but few (and appropriate) strategic and operational limitations. Success was never assured; we took many risks during both our preparation and conduct of operations. Agencies and coalition nations did not always agree on approaches and methods, but never did interference by "higher" impede the mission. We were free to learn, adapt, and compromise as necessary, accruing successes while retaining the respect and admiration of the local people.

Directive control was sometimes applied deliberately. On other occasions the span of command or distances involved precluded direct control.[29] Either way, I benefitted from clear strategic intent and freedom of action and the Solomon Islands benefitted from a capable and agile force sure in its capabilities and purpose. The people of the Solomon Islands were given a second chance for a peaceful and prosperous future. And, despite experiencing pressures of command and uncertainty beyond that normally placed on a battalion commander, I relished the opportunity and have attempted to apply the same trust in subordinates ever since.

ENDNOTES

1   It is little known to even many Australians that the Solomon Islands shares air and sea borders with Australia and is within a few hours' flight of Brisbane. This became a concern after the September 11, 2001 attacks as a lawless Solomon Islands could become a departure point for similar short-warning aircraft attacks against Australia.

2   Map courtesy of the Australian National University CartoGIS, item no. 11-007_Solomon IS.png.

3   Australian Defence Force photograph by Corporal Sean Burton, Honiara, Solomon Islands, August 13, 2003, http://images.defence.gov.au/fotoweb/archives/5003-All%20Defence%20Imagery/DefenceImagery/ImageLibrary/3/JPAU13AUGo3SBo101a.jpg.info#c=%2Ffotoweb%2Farchives%2F5003-All%2520Defence%2520Imagery%2F%3Fq%3DJPAU13AUGo3SBo101a.jpg (accessed January 7, 2016).

4   General Cosgrove is regarded as one of our nations most gifted communicators. This direct and concise piece of verbal guidance left me in no doubt as to the importance of the task entrusted to me and the consequence of failure!

5   Although I was now assigned to a joint operational level HQ, I remained dependent on my brigade HQ for some administrative support and was required (by my brigade commander) to back-brief my plans to him and to our higher Land HQ prior to formal presentation to HQAST for approval (a process not technically required but which was prudent).

6   The CIMIC capability had been highly valued during Australia's involvement in Vietnam but was not maintained in the ADF in the years after that conflict. Cosgrove had

resurrected the capability as a result of his experience as Commander, INTERFET in 1999 East Timor. He considered CIMIC essential to RAMSI but, paradoxically, with an aid component already committed to this mission (AUSAID), CIMIC became a complicating element rather than an important interface. I drew the element down to one liaison officer after a month in-theater driven in part by request from AUSAID to do so.

7   As it played out, an additional rifle company from NZ was deployed into theater one month after our arrival due to a shortage of infantry manpower. This vindicated my early concern.

8   Notably, with 2 RAR's battalion (-) deployment to the Solomon Islands, we became the first RAR battalion to be concurrently deployed in two operational theatres.

9   One notable case of where my intent was not met was among the CIMIC elements. I issued clear guidance that CIMIC were not to conduct assessments without my express permission as I did not want to create expectations of either military support or non-sustainable solutions. Despite this guidance, in a number of cases CIMIC operators appeared in villages to do assessments of electric generators, irrigation systems, or perform other activities without permission. This created resentment when the military did not then return to provide support even if it had not been promised. After one month, I reduced the CIMIC component from 30 personnel to two to reduce this risk and ease tension with our civilian aid coordinators.

10 No other officer in my battalion headquarters had experience in a CJTF headquarters, nor had any completed staff college-

level training. This made their ability to conceive a form of headquarters beyond a typical battle group problematic. My experience on exchange in the US proved serendipitous in this regard.

11 Artillery batteries are habitually assigned to Australian infantry battalions. The resulting familiarity better allows for establishing longstanding relationships; developing and understanding common tactics, techniques, and procedures (TTPs); and building inter-unit traditions.

12 A S-3 is a battalion or brigade-level operations staff officer. A J-3 is the operations officer for a joint staff, i.e., one in a command with more than one service component. S-3 and J-3 also represent the staff sections themselves in addition to the individuals heading those sections.

13 I replaced one task group commander of major rank and placed a company commander of the same rank on a formal performance warning during the deployment. The company commander (one of my battalion officers) remained in command while in the Solomon Islands but was ultimately relieved of his command after another incident back in Australia. I also returned one lieutenant troop (platoon equivalent) commander to Australia for a breach of rules regarding non-consumption of alcohol.

14 The Strategic Command Group is the CDF's senior committee of leadership in the ADF. Composition varies but it typically includes the CDF, Secretary of the Defence Department, the service chiefs, and head of the Defence Intelligence Agency.

15 The reconnaissance was conducted in civilian clothes as Australia had not yet received confirmation of backing

from the Solomon Islands Parliament, nor had Australia yet announced the pending mission. A covert visit was the only viable option as there would be no time for reconnaissance once announcements were made, but it was also essential that our presence did not assume such approvals ahead of the political debates in each country.

16 COL Paul Symon was Nick Warner's (the RAMSI Special Coordinator) military adviser. He was personally briefed by GEN Cosgrove (CDF) at the same time as I was. At that time the CDF made it clear that COL Symon was to act as his "strategic eyes and ears" and had no command authority over the RAMSI military force.

17 Map courtesy of the Australian National University CartoGIS, item no. 12-074_Guadalcanal2-01.jpg.

18 Australian Defence Force photograph by Warrant Officer Class Two Gary Ramage, August 25, 2003, http://images. defence.gov.au/fotoweb/archives/5003-All%20Defence%20 Imagery/?q=JPAU28JUL03GR057a.jpg (accessed January 7, 2016).

19 Australian Defence Force photograph by Warrant Officer Class Two Gary Ramage, August 25, 2003, http:// images.defence.gov.au/fotoweb/archives/5003-All%20 Defence%20Imagery/DefenceImagery/ImageLibrary/3/ JPAU25AUG03GR01008.jpg.info#c=%2Ffotoweb%2Farchives% 2F5003-All%2520Defence%2520Imagery%2F%3Fq%3DJPAU25 AUG03GR01008.jpg (accessed January 7, 2016).

20 Much has occurred to ensure better understanding and trust between agencies since this time. For example, ADF and AFP representatives now routinely attend each others' career courses and are incorporated in interagency training activities.

21   Noting the smaller span of command as the police contingent numbered only around 200 for the first months of the mission.

22   In the event of an inability to agree in-theater, we would pass decisions back to Australia for consideration by an Inter-departmental Committee (IDC). In one memorable case, the IDC referred our own conundrum back to us for decision! We were certainly not victims of micro-management.

23   These measures were explained to soldiers and police alike. While well understood at the command level, they were observed with mixed levels of enthusiasm at the tactical level and required constant supervision to ensure we achieved the desired strategic effect (the policy regarding local shopping being one such case).

24   Destruction of weapons in remote locations was surprisingly difficult to facilitate as either oxy-acetylene cutting torches or generator powered grinders were required, these being difficult to transport.

25   We also had two highly capable short-landing and take-off DHC-4 Caribou aircraft. These played an important role in accessing remote jungle strips at ranges beyond those reachable by helicopter.

26   Australian Defence Force photograph by Corporal Sean Burton, Honiara, Solomon Islands, July 26, 2003, http://images.defence.gov.au/fotoweb/archives/5003-All%20Defence%20Imagery/DefenceImagery/ImageLibrary/3/JPAU26JUL03SB019.jpg.info#c=%2Ffotoweb%2Farchives%2F5003-All%2520Defence%2520Imagery%2F%3Fq%3DJPAU26JUL03SB019.jpg (accessed January 7, 2016).

27 In fact, I had briefed one course of action whereby we forcibly seized Harold Keke on meeting him. The Special Coordinator had discounted this option on the basis that despite the strategic importance of Keke, it could undermine the trust placed in RAMSI among the broader population.

28 Harold Keke continues to serve his life sentence in Rove Prison on the outskirts of Honiara.

29 Famously (and unbeknownst to me), there were fifty-four majors deployed during the initial phase of the mission. I had these culled after a month, but not before the soldiers were referring to RAMSI as "Recruit Another Major to the Solomon Islands" and a deck of cards *a la* the first Gulf War had been produced featuring all these officers. I doubt any other lieutenant colonel has had as many majors under his direct command on an operation. That deck of cards remains one of my prized possessions.

9

MISSION COMMAND IN IRAQ: THE AUSTRALIAN EXPERIENCE ON OPERATION CATALYST

Major General Anthony Rawlins, DSC, AM, Australian Army

As preceding chapters make clear, Australian Army application of mission command varies according to situational factors and individual personalities. Such was also true during Australia's deployment of forces to Iraq. This chapter focuses on contemporary application of mission command via the experiences of tactical commanders who deployed during the period 2006-2009 during Operation Catalyst, Australia's contribution to Iraqi reconstruction.

THE STRATEGIC CONTEXT

Historian John Blaxland observes that the Australian government's decision to send forces to Iraq was carefully calibrated, evolving from a series of incremental commitments that preceded the coalition invasion in 2003.[1] This decision by Prime Minister John Howard was controversial.[2] Howard justified it on the basis of ongoing breaches of UN resolutions, yet many in Australia remained sceptical that this was his true motive.

On March 8, 2003, Howard introduced a motion in Parliament condemning Iraq's refusal to abide by these United Nations Security Council resolutions. Some days later he summarized the charged political atmosphere as follows

> The issue of Iraq is challenging, difficult, and perplexing. It is an issue that I know has produced divided responses not only in Australia but around the world. . . . We believe that it is very much in the national interest of Australia that Iraq have taken from her her chemical and biological weapons and [be] denied the possibility of ever having nuclear weapons.[3]

However, what resonated most strongly in light of the national debate was his overt expression of support for the United States

> Of course our alliance with the United States is also a factor, unapologetically so. America has given very strong leadership to the world on the issue of Iraq. . . . Alliances are two-way processes and our alliance with the United States is no exception and Australians should always remember that no nation is more important to our long-term security than that of the United States.[4]

Many commentators advocate that this motivation was Howard's principal determinant. His willingness to support the US in its post 9-11 fight against terrorism was, however, neither a blind nor naive policy decision. Nor was it unbounded. Historian Albert Palazzo argues

> Australia joined the war to advance its own policy objective: to improve its relationship with its great power protector. It achieved this goal with great skill and at very little cost, and showed that it is possible for a junior partner to advance its

strategic interests within a coalition dominated by a great power. *For Australia, what mattered most was not what was happening in Baghdad but in Washington.*[5]

National policy objectives drive military planning, force design, and the selection of operational areas. Palazzo suggests Australia's derivative military objectives supported this aim of reinforcing the Australia-US security relationship whilst seeking to minimise tactical risks, the latter reflecting a perceived societal sensitivity to casualties in a conflict characterised as a "war of choice."[6] He contends that this ultimately translated into close control of the military mission through successive Chiefs of the Defence Force (CDF), each attuned to Howard's direction to limit Australia's military objectives.[7]

Precise strategic calibration also influenced the capabilities deployed and national command and control architecture to oversee their application. The ability to maintain a tight rein on the approval of tactical missions was injected into the Australian national command system. Australia deployed a one-star level officer to command Joint Task Force 633 (JTF633) in Kuwait during October 2001.[8] He was to play an important role in ensuring maintenance of the government's national strategy, employing the colloquial "national red card" to preclude Australian participation in coalition operations judged likely to contravene Australian legal authorities or overstep Canberra's policy parameters.[9]

OPERATION CATALYST

Factors influencing Australian participation changed as the war transformed into a virulent counterinsurgency. By late 2004, international pressure for increased Australian participation was insistent. The US now sought broader contributions from the international community to train Iraqi Security Forces (ISF) in preparation for their assuming responsibility for security under a newly elected Iraqi government. On July 16, 2003, Australia initiated Operation Catalyst, the national contribution to the rehabilitation and reconstruction of Iraq.

Howard announced the deployment of a conventional ground force element, the *Al Muthanna Task Group* (AMTG), to Al Muthanna Province on February 22, 2005. Its mission was both to provide a stable and secure environment for the Japanese Reconstruction and Support Group (JRSG) undertaking humanitarian, engineering, and reconstruction tasks in southern Iraq and training Iraqi Army units.[10] Working in support of Iraq's interim government, the task group provided a tangible Australian contribution to multinational force efforts to develop a secure and stable Iraq.

The command and control arrangements for this deployment reflected the intricacies of coalition warfare. Whilst the CDF always exercised full command of all ADF elements at all times, theatre command for all deployed elements was delegated to the operational level commander, Commander Joint Operations (CJOPs), also located in Australia.[11] CJOPs then assigned tactical elements under operational control (OPCON) of the relevant coalition headquarters, in this case, Multi-National Division Southeast (MND-SE).[12] However, as previously outlined, another layer of 'national assurance' or local control was imposed between MND-SE and the task group, known as 'national command', exercised by the senior Australian officer in theatre, COMD JTF-633.[13]

AMTG-1, under the command of Lieutenant-Colonel Roger Noble, established many operating characteristics retained during subsequent rotations. A battle group that eventually grew to around 450 personnel, AMTG-1 incorporated a regimental-sized headquarters, cavalry squadron equipped with Australian Light Armoured Vehicles, infantry combat team mounted in the Bushmaster Protected Mobility Vehicle, training team, and support elements. A diverse array of other specialist capabilities such as unmanned aerial vehicles (UAVs) and intelligence and electronic warfare teams were integrated into the force, as was an Australian government national agency liaison (or policy) officer. Under operational control of the British-led MND-SE headquartered in Basrah, AMTG-1 operated out of Camp Smitty near the provincial capital of As Samawah.

Map 9-1: Iraq[14]

AMTG-1 handed over to Lieutenant-Colonel Peter Short's AMTG-2 in October 2005. AMTG-2 benefited from a more formal training regime instituted to expedite training of subsequent task groups. Key task group representatives were now also able to conduct an in-theatre reconnaissance prior to deployment. The opportunity informed task group planning and design of its mission orders.

AMTG-2 transitioned to AMTG-3 (under the command of Lieutenant-Colonel Michael Mahy) in November 2005. AMTG-3 subsequently relocated to Ali Air Base, Talil in the neighbouring Dhi Qar Province when the Japanese withdrew as local areas began to transition to Provincial Iraqi Control (PIC) from July 2006. The task group retained operational overwatch of the province while continuing to train Iraqi security forces under a capstone program known as Mentor, Monitor and Train (M2T). Australia assumed operational overwatch of of Dhi Qar Province when it transitioned to PIC on September 21, 2006. AMTG-3 was redesignated Overwatch Battle Group West (OBGW-1) in recognition of the expanded mission.

Operations nonetheless remained focussed on developing ISF capacity and providing support if the security situation degenerated beyond Iraqi forces' capacity. Provision of combat support was, however, largely a fictional construct. Activation required a complex and lengthy series of approvals through local and national Iraqi governments, the coalition, and eventually the Australian government. It was made clear to battle group commanders that such approval would come only after formal application from the government of Iraq to the Australian government in conjunction with demonstrative evidence proving the situation outstripped the capacity of indigenous security resources.

Thus the bulk of OBG(W) activity focussed on concurrent activities including engagement with key local leaders, collective and individual training of provincial leaders and Iraqi Army elements, coalition force and installation security, and intermittent support to the local provincial reconstruction team. These efforts were accurately characterised as "preventing the insurgent's cause from gaining purchase in the prevailing society rather than on combat operations against the enemy."[15] Yet many felt tactical restrictions imposed on task force freedom of action impacted Australia's reputation and soldier morale. Blaxland charts that "the net effect of this government-driven tactical approach was the absolute minimisation of Australian casualties. But this approach came at a price in terms of credibility with Australia's allies and coalition partners and soldier's morale."[16]

OBG(W)-2 deployed in November 2006. I commanded this rotation as Commanding Officer (CO) of the 2nd Cavalry Regiment. Our activities largely mirrored those of our forebears. OBG(W)-3 (under Lieutenant-Colonel Justin Ellwood) followed in June 2007. OBG(W)-4 under Lieutenant Colonel Chris Websdane deployed in November 2007. Operation Catalyst ceased when the final ADF elements returned to Australia in July 2008.

THE AUSTRALIAN PHILOSOPHY OF MISSION COMMAND

Mission command was not yet formally entrenched as Australia's joint command philosophy during Operation Catalyst. Nevertheless, its benefits as an operational command and control construct were well established within the Australian Army. Army doctrine at that time specified that subordinates were to be given a clear indication of their commander's intent—the result required, the task, resources, and any applicable constraints—but the subordinate commander was to be afforded freedom to decide how to achieve the required result.[17]

Equivalent doctrine and practice were also operative amongst various militaries in Iraq, most notably the US and British contingents. Australia's conceptualization of mission command had been closely modelled on equivalent US and UK doctrine and was therefore synchronous with Multi-National Forces—Iraq (MNF-I) and MND-SE command approaches. A far cry from the situation confronted by Australians under MacArthur in World War II, MNF-I and MND-SE orders were developed and executed based on a familiar command philosophy. AMTG and OBG(W) commanders unanimously confirmed that mission command was also empowered during the conduct of their pre-deployment training. Following the original AMTG-1 deployment, MNF-I and MND-SE operational orders were made available to deploying forces. Whilst some caveats on issues such as national rules of engagement and compliance with international and domestic legal considerations existed, no task or battle group commander could recall any specific written operational caveats placed on missions assigned them. These factors suggest that the essential precursors for the practical application of mission command were in place for Australian forces. The accepted philosophy for command and control in Iraq at the operational and tactical levels was clearly mission command.[18]

Tactical commanders similarly confirmed they had no recollection of an instance in which a mission command philosophy was inappropriate for Iraq. That is, there was never a suggestion that mission command was only relevant to high-end conventional warfighting. In fact, its pointed relevance to Operation Catalyst inspired CO AMTG-1 Roger Noble to publish an article stating

the key to effective, focused action is mission command. The philosophy of mission command must be believed and nurtured. To be effective, it must be built on the intellectual components of clear intent, trust, and accountability. The central moral component is trust. A physical framework must also be established to support decision-makers at every level, especially those in the midst of chaos and in close contact with the adversary.[19]

Despite this seemingly universal relevance, there was significant diversity in its practical application across Australia's national command in Iraq. The Australian experience highlights today's challenges in fully enabling mission command as the framework providing connective tissue between the strategic, operational, and tactical levels of war. That experience is more generally reflective of issues faced by Western military forces on modern operations that tend to be highly politicized and conducted within a coalition architecture.

CORRELATING AUSTRALIA'S STRATEGIC SUCCESS AND MISSION COMMAND

In his article "The Making of Strategy and the Junior Coalition Partner: Australia and the 2003 Iraq War", Palazzo argues that the Iraq campaign was a master stroke in Australian policymaking. His thesis is interesting in that some elements appear inimical to the traditional military view that strategic success relies on successful actions in a wartime theatre of operations.[20] Palazzo posits a somewhat antithetical hypothesis, arguing that military success based on articulated coalition military objectives was neither the principal aim nor ultimate determinant of Australian strategic success during the Iraq War

> Unusually, strategic calculation was at the forefront of the Australian government's senior political leaders and their military advisors. The Australian government of Prime Minister John Howard saw the War in Iraq as an opportunity to advance a long-held security objective, *one that had little to do with events in the Middle East*. For Australia, the policy goal for its participation in the Iraq War was the opportunity to enhance its relationship with the United States.[21]

Palazzo suggests tactical victories were neither necessary nor encouraged by Australia's strategic leaders. The mere presence of

Australian forces was sufficient to achieve the desired strategic objectives provided their activities were constrained to ensure that tactical defeats were avoided. Tactical action therefore needed to be closely controlled to ensure a visible presence whilst minimizing risk to the force. This careful balancing act necessitated clear unity of purpose in terms of the objectives to be achieved and risks to be mitigated.[22]

Further, and in palpable contradistinction to the operative tenets of mission command, Palazzo contends

Australian political-military divisiveness was not evident in the Iraq War. Howard and his senior general, the Chief of Defence Force General Peter Cosgrove (and later Air Chief Marshal Angus Houston) acted as one in regards to Iraq. Cosgrove understood the government's purpose and worked towards that goal. To keep the ADF on target, the CDF tightly controlled the mission and kept the Prime Minister informed of its progress. Contemporary military theory contains numerous references to the effect of the "strategic corporal." In Iraq the influence of the junior ranks was minimal as Cosgrove aspired to be the "tactical general." Throughout the Iraq War no issue was too unimportant for the CDF's strategic-level oversight. The commander of Australia's headquarters in the Middle East also served as Cosgrove's strategic-level theatre representative. He had direct access to the CDF—outside the formal chain of command—and kept Cosgrove alert to all activities across the coalition that might have an effect upon Australia's ability to secure its goals.[23]

This observation suggests a strong measure of strategic micromanagement in Australia's tactical operations in Iraq, something generally considered antithetical to the tenets of mission command. It is against this apparently dissonant backdrop that appraisal of mission command application in Iraq proceeds.

APPLICATION OF THE TENETS OF MISSION COMMAND IN IRAQ

Australian Defence Doctrine Publication (ADDP) 00.1 and *Land Warfare Doctrine (LWD) 0-0* are both typical Western military doctrine in that they restate most of the universally-accepted prerequisites for the successful application of mission command. These include the concepts of reliability, trust, understanding, and risk.[24]

Given Palazzo's assertion of strong politico-military unity leading to invasive strategic control at the tactical level, I sought out the views of Australia's tactical commanders in order to understand how mission command was applied in practice. What follows are select snapshots from their experiences. It seeks to understand whether the Australian philosophy of mission command was applied in Iraq and, if effectively applied, whether it contributed to the successful strategic outcome cited by Palazzo.

Articulating Intent

Australian military doctrine suggests that the foundational basis for mission command will be absent without a clear articulation of intent. ADF doctrine therefore directs commanders to ensure that subordinate commanders understand the higher commander's intent, their own missions, and the operational context. They are to be told what effect they are to achieve and the reason why it is necessary.[25] Strategic commanders should specify the *ends* and allocate the *means*, but the *ways* should be left to the commander on the ground provided certain enabling preconditions are met.

The experiences of Australia's tactical commanders in Iraq suggest that these basic doctrinal stipulations were often only partially enacted. When they were, application was often ad hoc and informal. CO AMTG-1 recalls the strategic intent for his mission was verbally communicated by both the CDF and Chief of Army (CA) prior to his deployment. However, written versions of Australian strategic and operational orders were neither drafted nor made available until well after arrival in Iraq. Conversely, operational orders provided by the higher coalition headquarters, MND(SE), were precise and well articulated with clear intent provided in written and verbal formats.

Given that this was the inaugural—and relatively short notice—deployment, an element of "catchup" could be assumed and forgiven. However, subsequent commanders do not recall the same level of pre-deployment interaction with the CDF, CA, or CJOPs. Yet most tactical commanders expressed some level of confidence that strategic intent was sufficiently apparent even if this did not occur until the force had deployed. CO AMTG-3/OBG(W)-1 recalls

> after reviewing the intent that I drafted as part of our in-theatre review (completed at the end of the first month in theatre), it was pretty clear that I understood the strategic aim.... In terms of Australian interests it was clear to me that we had an "enabling role" and that the confluence between tactical mission and Australian interests lay in us providing an "overt demonstration of relevance and the continual delivery of positive outcomes or effects." The mission was on the nose politically.... It was clear that at the political-strategic level our masters were searching for answers and leaning very heavily on the military to come up with a solution. This was clear across the board within the coalition chain of command. I knew I wasn't going to get any coherent guidance and decided to embark on our own approach building on the conceptual work done by AMTG-2.[26]

Other commanders cite far lower levels of confidence in their ability to discern strategic intent in light of the political context. This was my own personal experience. Following pre-deployment training, I particularly wanted to confirm the truth behind the rumours that we were to avoid casualties.

Following the mission rehearsal activity, I sought out confirmation of strategic guidance, in particular an express articulation of any national caveats based upon the political situation, through a visit to Military Strategic Commitments in Canberra. In speaking to the relevant desk officer and asking what I needed to do and what I needed to avoid, the response was disappointingly blithe and vague—I was literally told "just don't f*#k it up." When I asked what this actually meant, I was told that I would very quickly discover what I was doing wrong if I "skied off-piste." But the desk officer couldn't cogently articulate what activities or actions were either "on" or "off-piste." But by the same token I was relatively comfortable that the pre-deployment training had provided me with an operative understanding of what we were there to do and what tools and tactics were available to me as a commander in seeking to execute this amorphous "national intent" in the absence of clear national orders.[27]

Other tactical commanders recount similar difficulties in determining a clear strategic intent during their mission analysis process. The provision of coalition operational orders during pre-deployment training provided a solid understanding of *coalition operational intent*. This did not, however, necessarily confer a precise understanding of *Australian strategic intent*. Australia's national orders tended to focus almost exclusively on the administrative or procedural aspects of constituting, training, and deploying the force. Some reassurance could be derived from the informal assertion that MND(SE) orders had been vetted at the Australian theatre command level and that instructions to overcome any incongruence would be clearly articulated in individual directives, HQ JTF633 national orders, or CJOPs campaign plans. But it was more likely they would be addressed "on occurrence" by the national command chain in theater. Lack of a clear, cogent statement of national intent stood out as a deficiency in the Australian preparatory regime.

Emphasis on administrative rather than operational aspects in national guidance meant there was no formal articulation of national caveats or "red cards" evident to tactical commanders, at least not before meeting with the national commander in theater.

Surprisingly, though Australian tactical-level commanders found it difficult to discern national intent and caveats, there was a view that our higher coalition headquarters had quickly grasped both. MND(SE) had by the time of the AMTG-1 to AMTG-2 transition already determined that Australian strategic level risk aversion effectively constrained the tactical commander's freedom of action. CO AMTG-2 observes:

I asked and waited but never received orders. I just took Roger [Noble's] orders and typed a "2" over the "1'" and no one noticed. By the time I arrived in theatre, the Brits had given up on us as far as contributing to their plan. The British brigade commander in Basrah gave me his brigade operational orders and asked me to fill in the blanks for AMTG-2. No one ever took me aside and told me what the actual (not for public consumption) reason for us was for us being in Iraq. It certainly wasn't to defeat anti-Iraqi government forces and it was also evident that the Japanese weren't totally incapable of looking after themselves. I had a conversation with Chief of Army just before we left and confirmed the real reason, but it wasn't until day three on the ground in theatre when I realised that all we could do was stuff things up . . . as As Samawah was as good as it was ever going to get.[28]

Without doubt, these experiences connote significant deficiencies in the articulation of strategic intent down to the tactical level within the Australian chain of command. However, nothing in this contradicts Palazzo's view that Australia's strategic leaders had developed and sustained a high level of "strategic unity." It seems evident that whilst there was a clear understanding of intent at the military strategic level, it was either deliberately or unwittingly not conveyed through the operational to tactical level of command. Further, the articulation of national intent and military strategic objectives down to the tactical level seemed to degenerate through subsequent rotations. This may have been because the political and military strategic leadership remained confident that the initial parameters had been set by the AMTG-1 deployment and thus only a minimal ongoing dialogue was required to maintain a steady state. Irrespective, what resonates most strongly in the accounts of subsequent commanders is the procedural failure or deliberate failure to convey the strategic intent—the fundamental strategic context—to their level as the tenets of mission command would mandate as an essential precursor for its effective application in this highly politicized operation. This failure also likely impeded the ability of the in-theater Australian national commander to exercise effective national command. Reliance on implied (rather than specified) intent meant that consistency in interpretation suffered.

In one illuminating exchange between myself, COMD JTF633, and CJOPs during the latter's visit to OBG(W)-2, glaring differences in the interpretation of my authority to intervene to support coalition forces were evident. COMD JTF633 had previously instructed that I was not authorized to intervene to support coalition forces in contact without his express permission. It remained a point of contention between us throughout our deployment as I believed this was within my authority set given our pre-deployment training, mission orders, and understanding of previous task and battle group standing operating procedures. In response to my request to confirm what was within my authority set, CJOPs stated that not only could I intervene at my own discretion within my tactical area of responsibility, but also in support of British elements in the adjoining MND(SE) provinces of Basrah and Maysan. This revelation stunned both COMD JTF633 and myself as this was never contemplated as being within either of our authority sets. It seemed contrary to the unstated but implied national intent to avoid combat unless absolutely necessary and was also inconsistent with other imposed limitations on activity within our tactical area of responsibility.[29]

This dissonance between commanders at three levels of command (strategic, operational, and tactical) was—to me—a seminal example of deficiencies in the articulation of command intent through the Australian chain of command during the OBGW(2) rotation.

These varied experiences suggest that the articulation of intent—the doctrinally essential precursor for the effective application of mission command—was too frequently absent. Not bringing tactical commanders "into the tent" in terms of the strategic intent meant it would be necessary to constrain their freedom of action in other ways to ensure their pursuit of tactical outcomes would not threaten national political objectives.

CONSTRAINTS ON TACTICAL FREEDOM OF ACTION

Australian doctrine at the time asserted two essential components also inherent in mission command: (1) commanders using a minimum of control so as not to unnecessarily limit subordinates' freedom of action, and (2) subordinates deciding for themselves how best to achieve their missions.[30] Given the cited deficiencies in articulating strategic intent and the lack of any specified limitations on tactical freedom of action in national orders, what remains untested are Palazzo and Blaxland's assertions that limits on freedom of action were exerted through the national command chain, in particular the Australian national commander in HQ JTF633.

Tactical commanders , recalled a variety of experiences regarding whether the in-theater Australian national command element enabled or undermined mission command via close tactical supervision of the force. CO AMTG-1 recalled that the Australian colonel embedded in MND(SE) headquarters provided excellent advice and COMD JTF633 (CJTF633) was also generally very helpful. In stark contrast, CO AMTG-2 was scathing regarding Australia's operational level commanders in place at the time unnecessarily constraining his tactical freedom of action:

> When we did receive NATCOMD [national command] guidance it was ad hoc and always centered on stopping us from doing something. First example—the Brits had responsibility for providing security assistance to the Iraqis for the December 2006 elections. It was in all our interests to coordinate and assist in this regard. We were told via national command we had no role and were to stay out of it. I attended the British lead planning meetings and contributed what we could by just advising how we would conduct our normal security roles . . . but it was a professionally humiliating time for us. Another example—we received what was assessed by the intelligence agencies to be credible intelligence that we would be hit by an improvised explosive device on one of the routes within the tactical area we were responsible for. I briefed CJTF633 who agreed to support so we planned an operation to interdict it. CJOPs was briefed and subsequently denied us this action and insisted the British execute it despite the threat fitting within our remit and emanating from our area. I was mortified.... It was professionally humiliating and a part of my respect for our organization died that day.[31]

I recall similarly invasive experiences during my OBG(W)-2 deployment that caused me to frequently question why COMD

JTF633 felt he needed to intervene in my tactical decision-making. Within a week of us arriving, Jaysh-al-Mahdi launched an attack on the police headquarters in As Samawah when its occupants refused to accede to demands to release some prisoners. Surprisingly, and reassuringly, for the first time the local Iraqi police and army elements combined and cooperated to defeat this initial attack. Jaysh-Al-Mahdi then returned in numbers, bolstered by fighters from other provinces, and laid siege to the provincial headquarters. The situation became dire; the ISF were outgunned and outnumbered and the local commanders called us for combat support and an urgent resupply of ammunition.

I was fully aware of our restrictions on intervention, so we arranged for an aerial resupply of ammunition and I also moved the battle group en mass to a secure location in the desert just outside the capital. We were just posturing, essentially bluffing, to alleviate pressure on the ISF—even if only temporarily. I was fully aware that we could not intervene in the fight—the purpose of this move was to create a ruse to deceive the insurgents that we were in fact committing...purely in order to unsettle them and to bolster the confidence of the local Iraqi commander. When I informed the JTF commander I had done this later that night, he exploded and insisted I had intervened in express contravention of national and coalition orders. His assertion was that I did not have the authority to move or "step up" my battle group; this required his express permission. I was quite flabbergasted. If I didn't possess the command authority to tactically maneuver my own force, what was my role?... In order to try and understand this, I cycled through the full range of actions that we had been instructed on and trained for on the mission rehearsal exercise with him, including actions that previous battle groups had all employed. He rejected each and every one of them as being outside my authority, including even the options of aerial overpasses by coalition aircraft or the ability to survey the situation with our UAV. His interpretation of my role and authorities was diametrically opposed to my mission orders and all the other advice, pre-deployment training and counsel I had received and had been endorsed during our preparation.

CO AMTG-2 recalled some similar experiences with HQ JTF633 during his deployment, leading him to conclude that

> To me, JTF633 became a conduit for getting reports out and acting as the unwilling scout for Big Brother. I realized very early to play it "grey" and stay under their radar or invite the 10,000-km screw driver from Australia. I was lucky to have great individuals as respective CJTF633 commanders during my time as they understood and applied mission command as best they could. By keeping things low profile I could achieve a modicum of freedom of action. I knew what the real strategic drivers were for the deployment; I got it. There was no need for unhelpful meddling by Australians who had no tactical situational awareness. To this day, I can only shudder at the thought of the meddling and obstacles I would have experienced if the operation had been more kinetic.[32]

It seems that much depended on the personality of individual national commanders and their corresponding interpretation of their role in safeguarding Australia's strategic intent. Whilst many acted in a manner described as an impediment to mission command given their requirements to vet and approve tactical missions, others exercised more of a mentoring and enabling role.

These various experiences tend to undermine Palazzo's argument regarding strategic unity leading to strategic micromanagement. Most tellingly, no task or battle group commander recalls any explicit intervention by the CDF interfering personally with his tactical decision-making. The most deleterious interventions seemed to have been exercised through the national command element. Different behaviors by different national commanders tend to suggest either a lack of clear articulation of strategic intent and/or national caveats down through the national command chain to them or, alternatively, an unwillingness by some national commanders to simply loosen the reins and allow tactical commanders freedom to maneuver. What is clear is that one of the key factors influencing the scope of freedom of action allowed was the extent of trust between national and tactical commanders.

TRUST

Western mission command doctrines posit that the trust a senior commander has in a subordinate will—and should—influence the freedom of action granted the latter. In light of Palazzo's assertions regarding strategic supervision bordering on micromanagement, I asked tactical commanders whether the prerequisite level of trust was afforded them.

It is important to note that mission command requires mutual trust at all levels of command. Trust involves a two-way relationship: subordinates are trusted to carry out their missions within the bounds of the commander's intent and commanders are trusted to back a subordinate when decisions so made nonetheless result in less than ideal outcomes.

Once again, great variability in the experiences of task and battle group commanders was evident. CO AMTG-3/OBG(W)-1 recalls

Both Commander 1st Division and Commander 3 Brigade were clear in their discussion with me prior to deploying about the political sensitivities but neither were prescriptive in their guidance, showing great faith and trust. The brigade commander completed a reconnaissance concurrent to mine…. We debriefed him on our appreciation while heading home out of Kuwait. He gave guidance but knew from his previous experience at US Central Command (Forward) and also with

INTERFET [International Force in East Timor] that guidance would remain fluid and that if we didn't stay a bound ahead in our thinking then we would be caught flat-footed.[33]

Conversely, my relationship with the national commander in place at the time reflected a lack of trust. This was because, from my perspective, the caveats he was imposing on the battle group were counterintuitive to national directions and emphasis on force protection. The lack of a unifying national strategic intent clearly contributed to this.

By way of example, during one indirect fire (rocket) attack on the base, insurgents fired a second salvo from the exact same point of origin at the thirty minute mark, which was instructive in that (1) they knew our protocols (i.e., all clear signals came after thirty minutes), and (2) they were confident there was going to be no response from us against the point of origin after the first salvo. I felt the latter had come to pass because COMD JTF633 insisted that our job was not to seek engagement with the enemy nor was it to protect the base. If a response was required, his view was that another national contingent should do it . . . even though one of our approved mission tasks was to patrol suspected indirect fire sites within our designated area. I felt strongly that this direction was inimical to our own force protection and yet I was unable to convince him that aggressive patrolling was actually a defensive measure designed to protect ourselves and that I was not—as he suggested—"under the spell" of another national contingent. We resolved to deliberately understate our patrolling plans on the basis that to do otherwise would invite opposition and in our calculus, increase the risk and potentially cost lives. The lack of clear national intent, leading to different interpretations as to what were authorised to do, and what we should do, fatally undermined the contract of mutual trust between us.

These differences in perceived authorities and the national commander's stipulation that he was to vet all my tactical orders meant that by late in the deployment we were selectively providing redacted plans to him. I had agreed on some protocols with his senior operations officer by which sufficient information could be provided to inform him of our tactical activities, but without invoking his suspicion or violating his risk sensitivities. His interventions had led me to doubt his tactical acumen and we could barely discuss operational matters without an argument. I therefore chose to seek counsel on the merits of our proposed operations with my parent brigade commander back in Australia. The fact that his assessments invariably supported our tactical thinking was reassuring in one sense, but troubling in another...it again reinforced that there were vastly different interpretations of national strategic intent and derivative tactical imperatives.

CO AMTG-3/OBG(W)-1 had what could only be considered the opposite experience:

> National command worked as well as it could, and worked because they supported us and didn't interfere. The national commander was an excellent mentor and supporter. As I have said before, I put this down to his operational command experience.[34]

FINAL OBSERVATIONS

What then can we distil in terms of the Australian mission command experience during Operation Catalyst? Do Palazzo and Blaxland's assertions that strategic success was borne of tight politico-military unity, unwavering strategic control of tactical operations, and a seemingly invasive national command architecture stand up to scrutiny? Is mission command compatible with these conditions? If so, did its application materially contribute to strategic success in Iraq?

The most obvious deduction is that application of mission command is a function of individual command style. Some

Australian commanders were more adept at it whilst others struggled to enliven it.

Those tactical commanders who felt most confident in their authorities and empowerment to act were those who received unequivocal strategic intent statements from the chain of command. The fact that this articulation varied so extensively between deployments must be considered a deficiency in Australia's campaign command architecture. In most cases, a strategic leader's personally articulated intent reduced confusion and alleviated concerns that would otherwise build as various interpretations of the strategic intent percolated through command layers. Commanders struggled to adduce strategic intent in the absence of its clear conveyance, leading to wide variations in perceived caveats on the tactical commander's freedom of action.

Despite Palazzo's assertion, no tactical commander recalled the CDF tightly controlling tactical action himself. Strategic micromanagement is difficult to confirm or disprove because it was clear that some in-theater national commanders unduly interfered with tactical commanders' decision-making whereas others proved a valuable mentoring asset. This might have been a matter of personality, a by-product of differing interpretations of national strategic intent, or a combination of both.

It is axiomatic to observe that trust is essential to effective mission command. The Australian experience on Operation Catalyst reinforces that trust remains a purely personal contract. Where trust existed between the national and tactical commanders, the experience in the application of mission command was overwhelmingly positive. In other cases a lack of trust or mutual respect seriously impeded the effective implementation of mission command.

It is all but assured that every Australian commander who served in Iraq would proclaim himself a proponent and effective practitioner of mission command. Yet the experiences recounted by Australia's tactical commanders suggest effectiveness varied considerably based on personal attributes and experience, the timbre of personal relationships, shared understanding of national intent, and mutual trust.

Whilst strategic unity at the political and military levels probably existed, the transmission of intent down through the military chain of command was frequently too ad hoc or episodic to be effective. Strategic micromanagement of battle groups might never have been executed by CDF personally, but it clearly occurred in varying degrees via the national command chain.

These recounted experiences are insufficient to determine whether there was a deliberate intent to withhold the full detail of Australia's carefully calculated strategic intent from tactical commanders as Palazzo suggests. Debate amongst the tactical commanders as to whether we were subject to overarching strategic direction to avoid contact with the enemy and avoid casualties continues to this day.

The Australian experience on Operational Catalyst suggests that selective application of some but not all tenets of mission command leads to a suboptimal outcome. Intent must be articulated— consistently and clearly—from the strategic to the tactical level if it is to be effective. A shared and consistent understanding of context at all levels of war allows the tactical commander to understand and better accept constraints on his/her freedom of action. This shared understanding is clearly a touchstone facilitating the development and sustainment of trust. It remains to be seen whether this is possible during highly-politicized operations or where there may be reasons to keep aspects of the strategic intent restricted to the highest levels of political and military command.

Nothing of the Australian experience in Iraq suggests deficiency in Australian or Western doctrine. Rather, it was the patchy and inconsistent application of mission command tenets that undermined its effectiveness. Whilst assertions of Operation

Catalyst's strategic success, particularly in terms of Australian national objectives, may well be true, few of Australia's tactical commanders would suggest the operation constitutes a template for emulation in the future application of mission command on military operations.

ENDNOTES

1 John Blaxland, *The Australian Army From Whitlam to Howard* (Port Melbourne: Cambridge University Press, 2014), 218.

2 The Opposition Leader at the time, Simon Crean stated "Labor opposes your commitment to war. We will argue against it and we will call for the troops to be returned." House of Representatives debates, March 18, 2003, p. 12512, http://www.aph.gov.au/About_Parliament/Senate/Powers_ practice_n_procedures/pops/pop63/footnotes#co1f21 (accessed November 4, 2015)

3 John Howard, transcript of the speech delivered by the Prime Minister of Australia to the National Press Club, in The Great Hall, Parliament House, March 13, 2003, http:// australianpolitics.com/2003/03/13/john-howard-iraq-speech-npc.html (accessed November 4, 2015)

4 Ibid.

5 Albert Palazzo, 'We Went to War for ANZUS', 25 March 2013 (2.00pm), part of the debate on Iraq ten years on, *The Interpreter* (emphasis added), available at: http://www.lowyinterpreter.org/post/2013/03/25/We-went-to-Iraq-for-ANZUS.aspx

6 Molan argues "The purpose of conducting a mission of choice is not necessarily to win (although rhetoric may spin on this issue) but to show commitment." Jim Molan, 'Choice and Necessity in Australia's Way of War', 2 September 2009 (11.03am), available at: http://www.lowyinterpreter.org/post/2009/09/02/Choice-and-necessity-in-the-Australian-way-of-war.aspx

7   Blaxland, The Australian Army From Whitlam to Howard, 218.

8   Operation Slipper was the overarching ADF contribution
     to the International Coalition against Terrorism. Operation
     Slipper, the International Coalition against terrorism,
     continued throughout the period of the war in Iraq.
     Operation Falconer, the ADF contribution to combat
     operations to enforce Iraq's compliance with its international
     obligations to disarm, commenced on 19 March 2003, ceasing
     on 16 July 2003 with the transition to Operation Catalyst,
     subsuming ADF participation in Iraqi security, national
     recovery and the transition to Iraqi self-government. See
     http://www.defence.gov.au/publications/lessons.pdf

9   Blaxland, The Australian Army From Whitlam to Howard, 233.

10   Australian Government, Department of Defence,
     Annual Report 2004-2005, http://www.defence.gov.au/
     AnnualReports/04-05/04_03_outcome1_09_spf.htm (accessed
     November 4, 2015)

11   Theatre command is the authority given by CDF to CJOPS
     to command assigned forces to prepare for and conduct
     operations (campaigns, operations, combined and joint
     exercises, and other activities as directed). Australian Defence
     Doctrine Publication 00.1, Command and Control, Canberra,
     Defence Publishing Service, 27 May 2009, 3-4.

12   Operational control is the authority delegated to a
     commander to direct forces assigned so that the commander
     may accomplish specific missions or tasks which are usually
     limited by function, time or location; deploy units concerned
     and retain or delegate tactical control of those units. ADDP
     00.1, 3-10.

13 National command is a standing command authority conferred upon a national appointee to safeguard Australian national interests in combined or coalition operations. National command does not in itself include any operational command authorities. ADDP 00.1, 3-5.

14 Map courtesy of the University of Texas libraries, The University of Texas at Austin, https://www.lib.utexas.edu/maps/middle_east_and_asia/iraq_pol-2009.pdf (accessed December 23, 2015).

15 Blaxland, The Australian Army From Whitlam to Howard, 239.

16 Blaxland, The Australian Army From Whitlam to Howard, 242.

17 Land Warfare Doctrine 0.0, *Command, Leadership and Management*, Sydney, Headquarters Training Command, November 17, 2003).

18 ADDP 00.1, 2-8.

19 Roger Noble, "The Essential Thing: Mission Command and its Practical Application," *Australian Army Journal* III, (Summer 2006): 124.

20 The mission command philosophy provides the tactical commander with the strategic and operational context needed to inform his tactical action. Through feedback, the tactical commander provides the higher commander with the situational awareness needed to enable adjustment and recalibration of higher echelons' direction.

21 Albert Palazzo, 'The Making of Strategy and the Junior Coalition Partner: Australia and the 2003 Iraq War', *Infinity Journal* 2 (Fall 2012): 27 (emphasis added)

22 Palazzo cites specific control factors that enabled Australia
    to implement a calculated military plan based on a strategic
    policy objective distinct to the tactical requirements of the
    coalition leader. Australia's key strategic leaders intuitively
    understood that: (1) all of Australia's key military decision-
    makers needed to understand the policy objective and remain
    unified in its pursuit, (2) the force structure of Australia's
    contribution, its capabilities, and risk mitigation measures
    needed to be defined early and then adhered to, and (3)
    the force elements needed to be capable of performing the
    assigned tasks (this being more important than the size of the
    contribution). Palazzo, "The Making of Strategy," 28.

23 Palazzo, "The Making of Strategy," 28.

24 ADDP 00.1, 2-9 to 2-10; and LWD 0-0, 2-4 to 2-5.

25 LWD 0-0, 2-4.

26 Extract from author email received from colleague.

27 Extract from author email to colleague.

28 Extract from author email received from colleague.

29 For example, OBG(W)-2 was heavily constrained in that we
    could not move north of the Euphrates River in Dhi Qar
    without Minister of Defence consent. This was because Dhi
    Qar was considered to be a highly dangerous province, and
    this artificial constraint was seen as deterring the battle group
    from operating in the more dangerous areas of the province.
    However, an arrangement existed between Australian and
    British forces whereby mutual support could be provided
    between provinces where the respective nation's forces were
    in heavy contact and in danger of tactical defeat. In order

for the battle group to be able to rely upon support from the British, it needed to be seen to be responsive to requests from British forces in Basrah and Maysan provinces. This determination that a decision to deploy to another province to support British forces in heavy contact was within the battle group commander's authority must be seen as highly dubious, given the aforementioned restrictions placed upon operations north of the Euphrates River in Dhi Qar province, and particularly given the constant rate of combat being experienced by British forces in Basrah and Maysan provinces.

30 LWD 0-0, 2-4.

31 Extract from author email received from colleague.

32 Extract from author email received from colleague.

33 Extract from author email received from colleague.

34 Extract from author email received from colleague.

10

Mission Command and the 2RAR Battle Group in Afghanistan: A Case Study in the Relationship between Mission Command and Responsibility

Brigadier Chris R. Smith, DSC, CSC, Australian Army

There are two freedoms; the false, where man is free to do what he likes; the true, where man is free to do what he ought.

Charles Kingsley

I was the commanding officer of the 2nd Battalion, The Royal Australian Regiment (2RAR) in late March 2011. Less than three months later, the battalion would be in Afghanistan. Some 350 officers and soldiers from a range of other units and brigades augmented the battalion to form a battle group. We were in the early stages of pre-deployment training when I received a visit that would have important consequences for the remaining duration of my command.

A colleague who had recently returned from Afghanistan was so troubled by what he had discovered there that he wanted to make his findings known to me. He thought that his anecdotes might serve as a timely warning for the battle group. The information he revealed to me and my regimental sergeant major, Warrant Officer Class One John Pickett, had a profound effect on both of us.

The visitor placed a disk into my computer and showed me sections of video taken by soldiers in Afghanistan. The video showed irresponsible, careless, and unprofessional behavior.

In one instance, the video showed the members of a patrol on a high feature overlooking homes on the valley floor below and well within the range of small arms fire from those buildings. The patrol was in a contested area and its members were well aware of the presence of Taliban fighters.

The patrol's disposition could only be described as a holiday atmosphere. Soldiers and their commanders were standing or sitting in foldout chairs. They listened to music and coalesced in tight groups, presenting a lucrative target. They were facing each other, conversing loudly, inattentive to their surrounds. Most did not have weapons within arm's reach. Few wore helmets or body armor.

The visitor then showed me other similar examples. One series of photos showed about ten men standing shoulder to shoulder in a semicircle. They were watching another soldier attempting to dig a very large boulder out of the ground for the juvenile thrill of seeing it roll down the slope to the valley floor below. It seemed that the soldiers were standing about twenty to thirty meters outside the perimeter of their position. Again, most did not have weapons within arm's reach. Few wore helmets or armor.

In addition to the visitor's vignettes, I had recently become aware of a tendency for some small units to set up weights and other gym equipment in tactical overwatch positions.

These examples alone, while disconcerting, might easily be dismissed as inevitable and probably infrequent occurrences of indiscipline; *twas ever thus*. Australian soldiers, like all soldiers I suspect, will do what they are allowed to get away with, sometimes to their detriment. Even under the best of commanders, soldiers will act up. But what troubled my colleague the most was that although not universal, the array of images and video seemed to represent something more than just the odd instance of indiscipline.

In coming to me, my colleague's intention was to alert me to the potential for complacency and casual attitudes to develop in the battle group, and to remind me of the seriousness of the consequences of the same. He believed that the consequences of the casual attitudes represented in the images might have been tragic in at least one instance. While the regimental sergeant major and I were alert to the potential for complacency within the battle group, this visit prompted us to pay greater attention. We would be arrogant or naïve to think our battle group would be any different if these things were occurring in other units under highly capable and proven commanders.

This chapter opens with a quote from writer Charles Kingsley. Through the prism of my experience in charge of the 2RAR Battle Group during its 2011 tour of Afghanistan, I aim to demonstrate that mission command is "false" when commanders use their relative autonomy to do what they like and "true" when they use that autonomy to do what they ought. I aim to alert future commanders to the potential to confuse mission command for something more akin to "hands off" leadership, causing them to misapply it.

Mission command derives from an appreciation that warfare is dynamic, complex, and unpredictable. Any attempt to centrally direct actions of modern armies and their diverse sub-elements is self-evidently problematic. Armies controlled in this way tend to be unwieldy, sluggish, and therefore unable to take advantage of unexpected, fleeting opportunities or respond quickly when suddenly confronted by an unfavorable situation. The wiser approach is to directly control only as much as is necessary and no more, allowing subordinates the autonomy necessary to deal with changing circumstances as they see fit.

Yet the introductory anecdote suggests mission command is more than just the idea of a superior commander devising a mission and leaving its achievement to the leader on the ground - it is not "hands off" leadership. Failing to exercise appropriate supervision puts the mission and soldiers' lives at risk. There is thus an onus on the subordinate commander to act responsibly within the intent of his superior and for all commanders to exercise appropriate levels of supervision.

Following the example of Field Marshal William Slim's *Defeat into Victory*, the chapter is somewhat "warts and all." Without the warts, the chapter would have been so refined to amount to nothing more than an elaborate list of banal principles. Principles can only be properly understood when placed in context. My aim is to therefore provide that context, to forearm future leaders with an appreciation of the challenges of mission command in practice.

The remainder of the chapter has three parts. The first establishes the context of our battle group's operations, which is necessary to appreciate the particular way we practiced mission command as described in part two. The final part builds on those previous to consider specific examples of mission command in action and illustrate the relationship between mission command and responsibility.

AUSTRALIAN SUPPORT TO THE INTERNATIONAL SECURITY ASSISTANCE FORCE (ISAF) AND THE BATTLE GROUP'S OPERATIONS

The Australian government's officially stated objective for committing military and other capabilities to Afghanistan was to deny Al-Qaeda sanctuary. The war had lost much of its popular support in most contributing coalition countries by 2011. Many in the Western world had doubts about the merits of continued foreign intervention and were doubtful regarding the likelihood of its successful outcome. This international cynicism and growing discomfort amongst the US public likely contributed to the US government's December 1, 2009 announcement that ISAF would transition security responsibility to Afghan security forces by 2014. The long withdrawal would commence in June 2011, the month my battle group deployed to the theater.

Many analysts agreed that defeat of the Taliban, if achievable, was likely to take several years and was dependent on denial of sanctuary in Pakistan. The US government's timeframe for transition of security responsibility therefore seemed to signal acceptance that the international coalition would not defeat the insurgency before departure of most foreign troops.[1] If the Taliban were to be defeated, Afghan forces would have to do it.

Preparing Afghan institutions to assume responsibility for the Taliban's defeat therefore became the international coalition's primary mission. Australia's principal contribution to the war in 2011 was a battle group designed to bring the Afghan 4th Brigade, 205th Corps (4/205 Brigade) up to an enduring level of capability sufficient to underwrite the ultimate defeat of the Taliban after the departure of international troops.

The battle group faced a problem with two crucial elements in particular. We had to (1) assist 4/205 Brigade in becoming independently capable of making a credible contribution to its country's security, and (2) contain the insurgency in 4/205 Brigade's area of responsibility (Uruzgan Province) to enhance its chances of ultimate success. We deduced that the latter should follow by doing the former well. In other words, assisting in development of a more capable 4/205 Brigade would abet Taliban defeat.

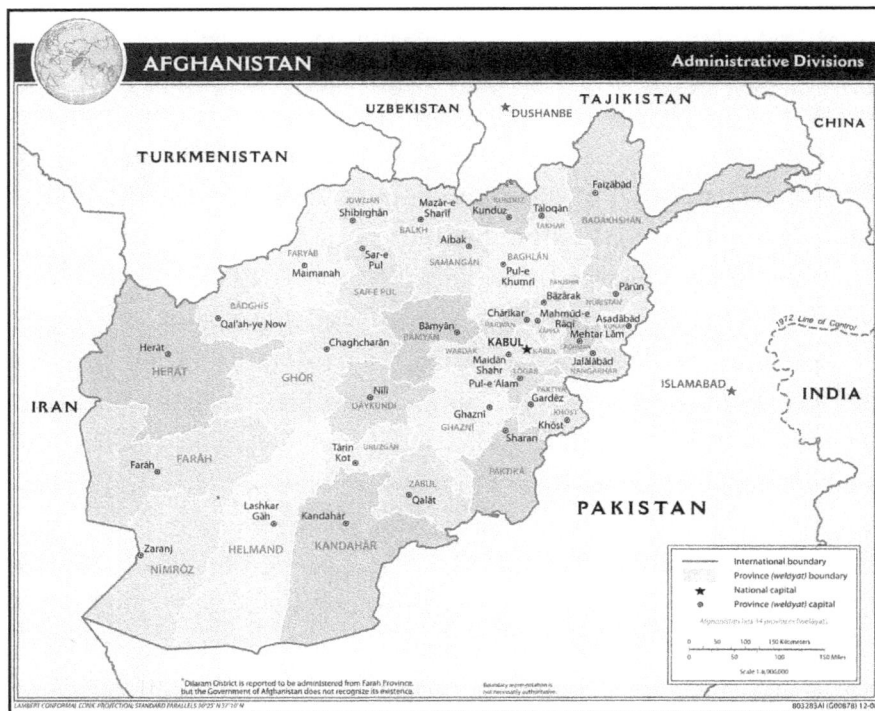

Map 10-1: Afghanistan Provinces[2]

The Taliban had two critical advantages over the counterinsurgents. The first was its willingness to use coercive violence to control the population. Fearing that withdrawal of foreign troops would result in the Taliban seizing control, many locals were prone to providing either direct or tacit support to the enemy, willingly or otherwise. The second advantage was the ability of Taliban fighters, informers, spies, and agents to hide in plain sight among the population.

People's attitudes and allegiances were extremely difficult to discern. Intractable tribal animosities and loyalties drove popular behavior as much as any other factor. It seemed that members of the population tended to collaborate with the side that exercised control most effectively irrespective of politics, capacity building efforts, or aid programs. It followed that we should enable 4/205 Brigade and national security forces more broadly to achieve control over as much of the province as they could.[3] A focus on control came to characterize our mentoring of 4/205 Brigade.

4/205 Brigade, choosing not to use coercion and intimidation to control the population as did its enemies, needed to remain amongst the population to prevent insurgent intimidation. Because there were too few troops to control everywhere, it was also important to put pressure on insurgents in their sanctuaries within Uruzgan Province. 4/205 bases were therefore both foundations for population support and launching points for actions against the foe. Imparting the enthusiasm and confidence necessary to mission success across the widely dispersed 4/205 Brigade became our primary focus.

ISAF directives dictated curtailment of direct coalition forces' involvement in security operations. My orders were therefore to continue withdrawing Australian mentoring teams away from smaller remote outposts as begun by our predecessors. These teams would relocate to several primary bases in the province that were also the headquarters for the 4/205 Brigade infantry *kandaks* (battalions).[4] By the middle of the tour almost all 2RAR Battle Group soldiers were therefore concentrated in five bases.

As our mentoring teams withdrew from the smaller outposts, we found that there was a consistent failure of the remote Afghan National Army infantry *tolays* (companies) to continue to act against the enemy. It was apparent that most would simply remain safely inside the walls of their patrol bases, venturing out only to resupply food or water. Given the essential requirement for frequent small local actions and forays into enemy-controlled areas, this lack of activity was an alarming indicator of possible long-term mission failure.

AUSTRALIAN APPLICATION OF MISSION COMMAND IN 2011 AFGHANISTAN

To address the failure of the 4/205 Brigade infantry tolays to act in our absence, we sent out mobile teams to all outposts as often as possible. Dispatching small teams under junior leaders to motivate Afghan officers and soldiers to seek out, kill, capture, or deter the enemy thus became our *modus operandi.*

This was the epitome of decentralized operations. It required that I give junior leaders a good deal of command autonomy. It necessitated the fullest expression of mission command because any attempt to direct the actions of the small patrols would stifle the autonomy necessary for taking advantage of unforeseen opportunities and responding quickly when suddenly confronted by unfavorable situations.

In order to make my intentions clear, I wrote a rather lengthy analysis of our mission and the factors bearing on it early in our pre-deployment preparations.[5] In addition to a thorough assessment of the situation, I discussed the character of insurgencies, including what might constitute success, how progress would be difficult to discern, and how our focus would be entirely on 4/205 Brigade. This document was the basis of officer training and I included it in the battle group's operation order.

I spoke to our soldiers as frequently as I could. I kept the logic simple but emphasized the direct connection between 4/205 Brigade's success and our own mission accomplishment. This message was particularly important because colleagues with experience in Afghanistan had warned me that there was a tendency for some Australian officers to prefer to operate independently of Afghan soldiers whenever possible.

The tendency was not surprising. Many young soldiers were particularly keen to come to grips with the enemy. Afghan soldiers would often make it difficult for them to do so. The behavior of Afghan soldiers could be dangerous whether in contact with the enemy or not. Further, Australian soldiers at times had difficulty accepting certain Afghan customs. We were without doubt asking a lot of our young leaders.

I continued these conversations with our battle group in Afghanistan. Like my predecessors, I travelled regularly to the remote companies and platoons. I used these occasions to appraise them of the broader situation, and gave the soldiers a sense of progress and how their hard work was having an effect. I made a point of describing progress in terms of 4/205 Brigade kandaks and tolays.

That my operations officer, Major Ben McLennan, was very capable allowed me to be away from the headquarters for extended periods to see to the proper execution of the mission. I trusted that he could plan operations and issue orders in my absence, particularly if communications were difficult. Moreover, because local circumstances rather than the broad mission tended to change, those orders pertained mainly to reorganizing or apportioning forces and equipment in response to shifts in the main effort.

The most important communications between my company commanders, principal staff officers, and me took place during nightly conference calls on the radio net or the secure telephone system. I expressed my intentions through these relatively unstructured discussions. The simple direction to imbue Afghan troops and their leaders with a lasting will and habit to win never changed, so conversations overwhelmingly regarded changes in the situation and subtle adjustments to that intent.

I encouraged input from my subordinates, so I tried to set a climate in which company commanders, the regimental sergeant major, principal staff, and I were unafraid to voice doubts, challenge each other, and adopt improved approaches. We updated each other not in accordance with a checklist but rather in keeping with our collective sense of what we were learning and what was changing. We questioned each other, our understanding of conditions, and our conjecture and assumptions. While the conference calls obviously included necessary coordination, I did not allow detailed instructions to impede important discussions about intent and mission context.

***

Yet, from the very beginning, my colleague's visit and the potential for complacency weighed heavily on my mind. I was convinced that casual attitudes were likely to thrive in the battle group if I left it unchecked. I had to somehow control only as much of the operations as was necessary while also ensuring that my officers and noncommissioned officers were leading responsibly.

Elaborating on the maxim that an organization does well only the things the boss checks, US General George S. Patton once observed,

> In carrying out a mission, the promulgation of an order represents not over ten percent of your responsibility. The remaining ninety percent consists in assuring through personal supervision on the ground, by yourself and your staff, proper and vigorous execution.[6]

The dispersed and decentralized character of the battle group's operations did not obviate my officers, noncommissioned officers, or my responsibility to see to the proper execution of my intentions. Nor did it obviate our responsibility to coach, mentor, and correct our subordinates. Like all commanders, I had to balance the somewhat competing requirements of granting a high degree of command autonomy and "assuring through personal supervision on the ground ... proper and vigorous execution." I concluded that frequent personal visits to patrol bases, accompanying troops on patrols, and positioning myself forward during larger-scale operations would be required.

I also tried to instil a belief throughout the battle group that small things mattered. I did so by talking about the importance of such issues as how the death of a mate in action would be difficult enough but how the death of a mate because of irresponsible, careless, or complacent conduct would be unbearable. I tried to imbue the men and women of the battle group with the belief that small things must be reinforced through personal example too. In addition to trying to set a good example myself, I also made a point of getting around and personally correcting soldiers for failures to do the small things well, which was not well received.

My company commanders and sergeants major by and large got the message early on. Before deploying into the theatre, I sensed that my efforts—and the efforts and example of my regimental sergeant major, company commanders, and company sergeants major—were having the desired effect on our officers and noncommissioned officers. But our efforts were inadequate. We did not get through to everyone.

One of my most significant failures was not being able to get my intent through to enough of the officers and noncommissioned officers and to the extent I believed was necessary, the reasons for which I explore later in the chapter. Whereas I felt I could afford to grant all my company commanders a similar degree of command autonomy, I may not have sufficiently accounted for their relative abilities to get subordinate leaders to lead responsibly. Some were more stringent than others, which may have led to an inconsistent approach across the battle group.

***

One of my company commanders left Afghanistan early to take up a new appointment in Australia. Major John Eccleston, his replacement (who had once been a platoon commander of mine when I was a company commander and in whom I had great confidence), quickly captured the flavor of mission command, including the dual characteristics of *intent with simple direction* and *overseeing responsible execution*:

Your guidance to me on arrival prior to taking command of

the company was simple and to the point. It articulated the problems I would face and the desired end state you wanted me to achieve. You discussed with me two key points. You warned me of the potential for complacency and you wanted me to motivate the kandak to conduct operations and dominate the area of operations. Essentially, you wanted me to address responsibility and get the Afghan soldiers motivated to own their area of operations. To me this was simple, but also a sign that you trusted me to get on with company-level mentoring operations within my area of responsibility.

My instructions to my engineer commander, Major Barry Mulligan and his sergeant major Warrant Officer Class Two Jeramie Faint early in the tour, provides an example of the difficult challenge of balancing the two characteristics of granting operational freedom to my commanders while ensuring that officers and noncommissioned officers were acting as they ought.

The training of engineers in improvised explosive device search and demolition was intense and of very high quality. It took place over a three-month period culminating in the South Australian desert at a place called Woomera. Noting the importance of this training and my responsibility to integrate the engineers into the battle group, I visited Woomera early in the battle group's pre-deployment training.

During my visit, the instructors told me that a number of the deaths and serious injuries to engineers dealing with improvised explosive devices were likely attributable to deviation from the techniques they were testing during this culminating exercise. They said that engineers in theater could become complacent, justifying unnecessary modifications to the proper techniques with the argument that they were adapting to changes in the situation. The instructors were unequivocal in their belief that "adaptation" was a euphemism for short cuts. I heeded the instructor's advice and urged Mulligan and Faint to maintain the standards set during the pre-deployment exercises.

Many leaders and soldiers received the resulting oversight poorly. Some engineers claimed that the supervision increased the stress on them, notwithstanding the fact that despite the discovery and demolition of hundreds of explosive devices we were fortunate not to have them kill or seriously wound a single engineer. While one may argue that this statistic is to some extent affected by luck, the season, and the type of operations conducted, it is a powerful measure of training quality, suitability of the techniques that were enforced throughout the tour, and—importantly—the oversight provided by the leaders in the battle group. Yet some junior leaders and soldiers mistakenly judged that the oversight was a consequence of a lack of trust in them.

The engineers were not alone in this regard. My emphasis on professional standards raised the ire of many others. I issued quite specific instructions about a range of issues related to responsible and attentive conduct. They included directions regarding readiness for battle, digging fighting positions during overnight halts, manning sentry positions with two soldiers rather than one, wearing of protective equipment, carriage of weapons, and prohibiting distracting and superficial activities such as desires to grow beards and wear unauthorized clothing.

Officers and noncommissioned officers resented these instructions, regarding them as micromanagement, contrary to mission command, and illustrating a lack of trust. They were to some extent right. My colleague's visit and the strength of the urging of the instructors at Woomera had made me wary. Trust is, after all, a psychological mechanism. The levels of trust between two people are a function of the nature of the relationship between them. Relationships take time to develop. I was only six months into my command when I deployed into the theater. Roughly half of my officers had been with me for less than three of those months.

I suspect many of my subordinates were initially similarly guarded regarding me for the same reasons.

I had two choices. I could either give the commanders the benefit of the doubt—assume their trustworthiness and let them go with a minimum of supervision—or assume that they might not yet be worthy of my trust and supervise them relatively closely until they proved otherwise. Given what was at stake and what I had learned from my colleague's visit, instructors at Woomera, and others, I thought it prudent to do the latter despite its being contrary to my natural tendency.

This anecdote provides an excellent example of the challenges inherent in actually practicing mission command. The engineers had the requisite expertise but my trust in the reliability of the engineer officers and noncommissioned officers was limited. I had to balance between two extremes of (1) unqualified trust and decentralization, and (2) highly centralized and near-constant on-site supervision. In this instance, I believed that without my prescriptions and Mulligan's and Faint's oversight my soldiers were vulnerable to the complacency and casual attitudes described in the introductory anecdote. While I expected that we would lose lives, I could not bear the thought of losing a life because of complacency, negligence, or as a result of activity that was superfluous to the mission. I did what I thought appropriate, balancing my familiarity with these leaders, knowledge of their expertise and relevant experience, and previously demonstrated reliability.

Mission command is not "fire and forget;" it is dynamic. One must constantly evaluate the character of the mission command practiced with the passage of time, taking the above factors into account. While I am not sure whether we ever found the right balance, we adjusted our approaches throughout the tour. Oversight generally diminished over time for engineer commanders, for example. Yet we increased our oversight of some leaders when evidence suggested there was the use of shortcuts or instances of complacency.

Some commanders proved their trustworthiness so entirely that they only needed oversight in the way a golf coach might correct subconscious bad habits accruing in a golfer's swing. Others proved to be more prone to inadequate leadership and required consistently higher levels of oversight.

I rarely doubted the collective competence and commitment of our battle group soldiers and their leaders. We were very well trained. Yet some subordinate commanders were not doing what they ought. I was concerned these men and women would endanger both our mission and soldiers' lives. Unit visits reinforced these doubts. The regimental sergeant major and I patrolled and harbored overnight with units led by weaker officers. We witnessed posting of single sentries, lax attitudes when sighting tactical positions, seeking comfort at the expense of vigilance, and taking predictable patrol routes among other shortfalls.[7]

The following considers three more examples from our battle group's tour that further illustrate the relationship between mission command, responsibility, and trust.

Mission Command, Responsibility, and Trust

Australian doctrine emphasizes trust in its discussion of mission command, asserting, "mission command requires reliability of response, where commanders must regard their superior's intentions as fundamental guidance and make the attainment of such guidance the underlying purpose of every action."[8] It goes on to say, "high demands must be made on the leadership qualities of subordinates, on their initiative and on their sense of responsibility to carry out assigned tasks."[9]

Trust is therefore a function of trustworthiness—being reliable. Notwithstanding the fact that most people probably tend to think that they are trustworthy and ought to be left alone to do as they see fit, I found that the presumption of trust was a dogma among a significant number of my officers and noncommissioned officers.

Take for example an incident on Christmas Day 2011. That afternoon I received news of a negligent discharge of a weapon by a member of a protected mobility vehicle section.[10] Apparently the discharged round missed another soldier by a matter of inches. The investigation revealed that the negligently discharged weapon was an unauthorized AK-47, not a weapon issued to Australian soldiers. It turned out that one of the drivers kept the weapon behind his driver's seat and had done so for weeks, perhaps months.

When I questioned the section commander about the incident and asked whether he was aware that this soldier kept a weapon behind his driver's seat, he asserted he was unaware of the weapon. I asked him whether he inspected his vehicles, to which he responded by saying unashamedly that he did not. He contended that he trusted his men and therefore did not need to check them. His explanation of the nature of trust was in a patronizing tone that suggested I was unable to appreciate trust's nature or its importance.

Rather than taking responsibility for his soldier's negligence, the commander felt that his first obligation was to support his subordinate. When I asked him whether he still trusted this particular soldier, he replied that he did despite the extraordinary display of irresponsibility. He went so far as to attempt to justify why the soldier had the unauthorized weapon in the first place.

The example was not unique. I learned that many of my officers and noncommissioned officers regarded inspections and checks—or any direct oversight of their subordinates for that matter—as unconscionable expressions of a lack of trust. It seemed as though many thought that they ought to give their subordinates unconditional trust irrespective of the subordinates' reliability or conscientiousness. And so it seemed that a significant number of the battle groups' leaders held a rather dogmatic view of trust that may have contributed to tolerance and ignorance of shoddy practices and casual attitudes.

My own failure to conduct more formal and frequent inspections almost certainly contributed to the habit. I had failed to establish a culture of checking and inspecting. Nonetheless, there seemed to be something more to it. I struggled to understand how so many leaders had come to perceive trust in such a dogmatic way that it served as a barrier to good leadership, responsible conduct, and—most importantly—effective mission command.

The dogmatic presumption of trust also manifested itself among some leaders as a sense that the autonomy inherent in mission command equated to an inviolable rule that the commander more familiar with local conditions should be allowed to do as they like and that their judgment and decisions were sacrosanct. A number of individuals railed at the nerve of their superiors to challenge the sacrosanct institution of the "man on the ground."

It was as though just being the commander on the ground was sufficient justification for one's choices, whatever they might be. For example, an officer attempted to justify his decision to disregard orders and expose his men to unnecessary risk of death or injury by undertaking a routine patrol with baseball caps rather than helmets on the basis that he was better familiar with local conditions and judged it appropriate. This view of the sanctity of the subordinate commander's judgment was in essence a way of avoiding being accountable for one's decisions. The perception was somewhat understandable. Our *modus operandi* meant that quite junior leaders were bearing the greatest share of the operational load. They were the one's working most intimately with the Afghan troops, leading them into the fray, and enabling them to prevail.

Reflecting on the mistaken presumption that the judgment and decisions of the local commander were inviolate, D Company's sergeant major, Warrant Officer Class Two Andrew Munn, wrote:

Asking why sections and platoons did or did not do certain

things occurred a few times during the deployment such as when an officer unilaterally withdrew his troops from a cut-off position compromising an entire operation. Although the commanders felt like their actions were being questioned, I saw it as an after-action-review and used the incidents to give them a more aggressive mindset and mission focus for the next incident. I saw first hand the results of our questioning. On one occasion, the lead vehicle in a patrol was attacked (low order mine and small arms fire). The vehicle I was in, which would have sat in a parade-like formation before the questioning, immediately broke from the sealed road and moved rapidly to a fire position on the high ground to a flank. Based on his previous experience, that crew commander (who had multiple tours) would not have reacted the same way before we had the opportunity to question his actions and assumptions.

The next example suggests that these distorted views of trust had some relationship to my failure to connect many of my subordinate leaders to the mission and my intent.

***

The response by one noncommissioned officer to an address I gave to my cavalry squadron (company) just prior to an operation illustrates an ambivalence or ignorance of the purpose of the battle group and mission context that existed in some parts of the battle group. The operation was to commence the next day and involved a large vehicle patrol to test the viability of a roughly thirty to forty kilometer route from the provincial capital of Tarin Kot to the district capital of Khas Uruzgan. We knew the route as Route Whale.

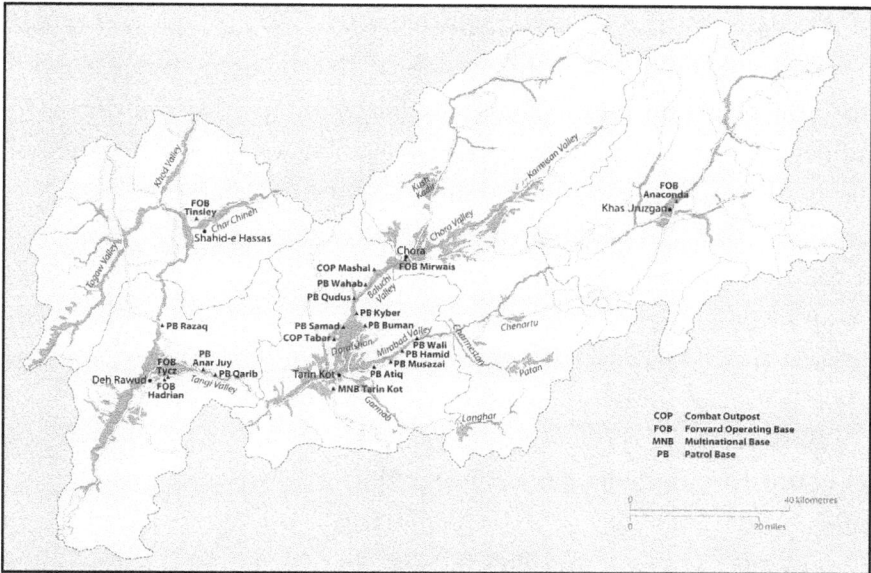

Map 10-2: Uruzgan Province, Afghanistan[11]

My address to the soldiers was largely an expression of my intent for the operation, focusing mainly on its rationale. I explained that there was a tolay of 4/205 Brigade in the Khas Uruzgan area that relied on ISAF helicopter resupply to sustain itself. Helicopter resupply was very likely to be impossible after the withdrawal of ISAF. It would therefore be necessary for 4/205 Brigade to sustain this remote unit by road.

The patrol's task was to escort a 4/205 Brigade reconnaissance group along Route Whale to prove not only that the brigade's vehicles could negotiate the route but that Afghan patrols and convoys could negotiate the route securely. In other words, the patrol was to establish the viability of the long-term sustainability of 4/205 Brigade troops in the important and contested Khas Uruzgan district.

Having explained the rationale for the task and its direct connection to the ISAF purpose, a noncommissioned officer smugly stated, "Sir, mates don't let mates drive on Route Whale" as if to say "this task is not worth the risk to me and the rest of us." I asked who else shared his concern. It was clear that most did not, but there was certainly a significant camp of supporters. On reflection, I came to realize that it was an example of ignorance of the battle group's mission and ambivalence to my intent.

It is worthwhile noting that this attitude was not widespread. It was, for example, in stark contrast to the attitude of the soldiers of another company who were upset at me because I had not chosen them for this juicy task, hinting at a connection between organizational culture, leadership, and mission command.

Reflecting on the ignorance of the battle group's purpose and its context by some leaders, John Eccleston wrote:

On return from my first ten-day patrol in the Tagaw Valley where we were shot at on a number of occasions, an officer commented to me that I could go home happy now that I had been in contact. I was surprised by the remark and had to ask a few questions to clarify what I was hearing and realized two things. Firstly, the company was somewhat ambivalent about its mission and secondly, being in Afghanistan was, for some, all about their individual war stories not the greater collective goal of the battle group.

***

Mission command unfettered by leader checks and other appropriate forms of supervision was a dogma in the minds of many in the battle group. It seemed as though some were confusing mission command with "hands off" leadership. The notion of trust had for some taken a distorted form. There were leaders who paid but lip service to the primacy of the mission. These factors suggest that mission command has become something of a cultural phenomenon and that cultural impediments to its proper practice have developed. It is useful to consider the causes of these impediments.

My own leadership is the first place to look. These distorted views almost certainly had some relationship to my failure to connect battle group officers and noncommissioned officers to the mission and my intent, which was a serious failing of mine. While I am confident that I did not instill a dogmatic notion of mission command and trust, I was disappointed in myself that I could neither connect many of these individuals to the mission nor fix the cultural impediments until after the tour.

I spent considerable time thinking about the phenomenon and decided to put my thoughts on paper about two-thirds of the way through our time in Afghanistan. I conjectured there were several causes for the state of affairs but that the primary cause was craven leadership. I found that some contemporary officers and noncommissioned officers tend to encourage relatively high levels of familiarity, identify too closely with their soldiers, and feel a strong pressure to please them. It seemed that the traditional military value priority of "mission, men, self" had changed to one of "men, mission, self." My intent, the mission, and mission performance appeared to count for less than how some thought they were perceived by subordinates, causing them to acquiesce too readily to their soldier's preferences, often at the expense of my intent and the mission.

The causes of craven leadership are many and probably include things like contemporary social trends and poor tutelage by more senior officers like me. Many of the causes are unlikely to be particular to the battle group and the matter deserves more attention than I can afford herein. Needless to say, addressing such leadership became the focus during the remainder of my command tenure.

One of my failings was that I woke up to these phenomena too late. I should have known from my experience as a platoon commander in Rwanda in 1995, for example, that there is an almost inevitable tendency for isolated small units to develop an "us-and-them" mindset and to begin to deviate from established norms. It is therefore necessary to bring small units back under the influence of a field grade officer (major or above) as often possible. I came to this realization too late, making the job of a number of my subordinates more difficult than it need to have been.

The above notwithstanding, mission command - indeed command in any form - requires strong leaders who do not require their subordinate's approval for validation. There was a relatively robust correlation between weaker leaders and casual attitudes, cynicism, and poor mission performance. The stronger the leader, the greater the autonomy a senior can afford to grant him or her. Weaker leaders, on the other hand, must be supervised more closely. My failure was that I was unable to imbue all of our battle group officers and noncommissioned officers with the maturity and sense of responsibility to do what they ought rather than what was popular or expedient, and—most importantly—to put the mission before their soldiers and themselves.

Meghan Fitzpatrick touches on this point in her earlier chapter on the Korean War. Reflecting on the requirement for leaders of strong character, she recalled Lieutenant Colonel Francis G Hassett stressing that:

> a battalion commander must select his company commanders carefully. He cannot risk a weak officer in these key appointments. The company commander is about the highest-level of close, direct personal contact with soldiers. It is a big job for a young man, responsible at times for the lives of a large group of soldiers.[12]

While Hassett was referring particularly to company commanders, his point can be extended to all commanders who have close, direct personal contact with soldiers. It is a "big job" for a young officer—or noncommissioned officer—responsible at times for the lives of soldiers. They must be up to it.

***

In order to illustrate the relationship between responsibility, trust, supervision, and mission command I have focused primarily on negative examples and the effect of cultural impediments. My approach is not intended to imply that things were all bad, nor that all manning the battle group's leadership positions were irresponsible and did not do as they ought. On the contrary, most did. Complacency and casual attitudes were apparent primarily in certain pockets of the battle group, corresponding, as I said, with the strength of individual leaders.

Mission command does not just happen; time, opportunity, and patience are necessary to develop a culture in which effective mission command can flourish. Crucially, some officers and noncommissioned officers who started the year doing as they *liked* were doing as they *ought* by the end of the tour as illustrated in the next example.

The officer commanding A Company, Major Tony Bennett, deployed his youngest platoon commander to a remote tolay very early in the tour. Sharing my concern regarding the potential for irresponsible leadership, Bennett and his sergeant major visited the platoon to oversee its performance.

Bennett found that the platoon's leaders had a poor understanding of the mission and my intent. He sensed that the platoon, like others, had a strong desire to seek contact with the enemy for its own sake:

I recall listening in on the platoon commander's orders the evening before heading out on a patrol. It was a strangely collaborative set of orders, dominated by the noncommissioned officers. A couple of days earlier, his team had been in a small engagement with the enemy and the Afghan soldiers had quickly withdrawn. Expecting the platoon commander to exploit the opportunity to take the Afghan soldiers back into the area to restore their confidence and exercise control, I was disappointed to hear him tell his platoon that they (specifically the Australians) would be in the lead with the Afghan soldiers simply following up. The platoon commander also emphasized getting attack helicopters on station quickly in the event of contact, which would have only reinforced the Afghan soldiers' sense that capabilities that would not be available to them after ISAF's withdrawal were necessary to prevail over local insurgents. After the orders, I corrected him and reminded the platoon of the reasons why they were going on patrol - to motivate Afghan officers and soldiers to seek out, kill, capture, or deter the enemy for themselves, not do it for them.

An experienced and popular noncommissioned officer with previous experience in Afghanistan approached the company sergeant major after the orders. Reflecting the dogmatic view of mission command and an ignorance of the mission, he questioned why the company commander felt it necessary to check on the platoon and expressed his opinion that it was unwise to go back into the area from which the Taliban had just repelled them. He thought that Bennett was micromanaging the platoon and cynically suggested that Bennett was there just to "get in a contact" (something I would be accused of too).

The patrol went ahead the following day without incident; however, the experience caused Bennett to bring the platoon commander back to the company headquarters to enable him to supervise the officer more closely. Bennett's aim was to improve the young officer's understanding of the mission and to teach him what he ought to be doing when unsupervised.

The platoon commander responded well to his company commander's tutelage. Bennett grew to trust him as he became more reliable and was able to exercise a less familiar leadership style. Reflecting on the events, Bennett wrote:

Following the *green on blue*[13] attack in November, I chose this officer to lead the mentoring team to work with the Afghan soldiers in the search for the rogue soldier and maintain the confidence of the Afghan soldiers in that area. Although he continued to make some mistakes, which is to be expected, it was clear that he was capable of doing what he ought within my intent. My orders to him were therefore very brief the night of the attack—"Mentor the Afghan soldiers in the pursuit of the rogue soldier and make sure they remain effective at the check point."

With no planned end date, the platoon commander and his men remained in the area of the attack for over four weeks without finding the rogue Afghan soldier or making contact with the insurgents. Yet he continued to seek out opportunities to build the confidence of the Afghan officers and soldiers by enforcing a relatively vigorous program of activities despite the reluctance of the Afghan soldiers and the reluctance of his own soldiers who were understandably keen to find the rogue Afghan soldier themselves.

The platoon commander would have likely acquiesced to his soldiers' preferences earlier in the tour, but under the close supervision of his company commander he had learned to do what he ought rather than what his soldiers liked. When Bennett visited the platoon this time he found them to be disciplined and working very closely with their Afghan counterparts.

CONCLUSION

In this chapter I aimed to demonstrate that mission command is "false" when subordinate commanders use their command autonomy to do what they like and "true" when they use that autonomy to do what they ought. I have endeavored to show that the command autonomy inherent in mission command is qualified - it comes with enormous responsibility.

By retelling the story of the 2RAR Battle Group's operations, I hope that I have been able to establish that there are two essential and related mission command preconditions. The first relates to the subordinate commander who must be responsible and consistently do what he or she ought. The second relates to the superior, who must ensure that subordinate commanders fully appreciate the superior's intent and mission context, thereby enabling them to act responsibly. And I hope to have shown that establishing these preconditions is no trivial undertaking.

I believe responsibility is the essence of mission command. In fact, responsibility is probably more important to an army than mission command because mission command must follow responsibility, not the other way around. Where leaders are responsible, mission command is likely to follow as a matter of course. On the other hand, mission command is impossible if leaders are irresponsible.

If responsibility is a necessary condition for mission command practice, then mission command is by extension more a cultural phenomenon than command technique. If an army—or a unit for that matter—aspires to practice mission command effectively it must first imbue its officers and noncommissioned officers with a strong desire to accept responsibility and a habit to exercise it. My failing as a commanding officer was my inability to imbue a number of my leaders in this regard.

Notwithstanding these failings, history is likely to reflect on the battle group's tour positively. It was successful by most measures and the professionalism of many of its officers, non-commissioned officers, and soldiers was high. There was no loss of life or limb from carelessness or lack of vigilance. I suspect these results were to a considerable extent due to the level of supervision we exercised.

Yet mission command, like leadership, is contingent and dynamic. It is neither "fire and forget" nor is there but one type. It is the responsibility of the commander to ensure his or her subordinates understand that mission command requires discrimination in its application. The commander must consistently evaluate whether the degree of autonomy he or she grants a subordinate commander is appropriate. Lack of strong leadership that allows a drift to casual attitudes, a change of mission that takes a leader out of his or her comfort zone, and similar developments should all trigger the commander to consider adapting the type of guidance and level of supervision he or she employs in dealing with each subordinate leader *as an individual*. Using the same approach with all subordinates and in all circumstances fails to account for differences.

ENDNOTES

1   The United States was the coalition lead partner.

2   Map courtesy US Central Intelligence Agency and Library of Congress, http://www.loc.gov/item/2009575509/ (accessed January 18, 2016).

3   We assessed that this objective not only increased the likelihood of positive responses from the people. It also made it easier to identify the enemy because incentives for informing on insurgents and promises of protection from insurgent retaliation have much greater weight when the government can reasonably back up its claims to provide security and safety. The objective additionally allowed for more efficient application of assistance as an incentive to encourage attitudinal shifts in the long term.

4   The battle group also began mentoring an additional kandak from a major base in northern Kandahar Province in October 2011.

5   A practice common to Australian commanding officers at the time.

6   George S. Patton, Jr., *War as I Knew It*, (Houghton Mifflin, Boston, 1947), 308.

7   Single sentries are more likely to fall asleep or are more likely to be inattentive than a pair of sentries.

8   Land Warfare Doctrine 0-0, *Command, Leadership and Management*, (2008), 2-4.

9   Ibid.

10 A negligent discharge is the unintentional and careless firing of a weapon. A protected mobility vehicle is a large four-wheel armored troop carrier.

11 Map courtesy and with permission of the Australian War Memorial, https://www.awm.gov.au/exhibitions/afghanistan-australian-story/maps/ (accessed December 21, 2015).

12 Francis Hassett, "The Military Team," in *Korea Remembered: The RAN, ARA and RAAF in the Korean War of 1950-1953*, ed. by Maurice Bertram Pears and Frederick Kirkland, Doctrine Wing, Australian Army Combined Arms Training and Development Centre, 2002.

13 A green on blue attack was a colloquial term for a surprise attack by an Afghan soldier on ISAF soldiers. The motivations for these kinds of attacks varied, but included the perpetrator acting on behalf of the enemy. In this instance the attack had wounded three Australian soldiers and occurred just one week after another green on blue attack that killed three Australians, an interpreter, and wounded several others.

11

THE AUSTRALIAN SPECIAL FORCES APPROACH TO MISSION COMMAND

Brigadier Ian Langford, DSC and Bars, Australian Army

INTRODUCTION

As noted in the introductory chapter, *Auftragstaktik*'s ancestral roots date from the Prussian Army general staff in the late-19th century.[1] This philosophy of "leading by task" stood in stark contrast with the previous Prussian model of *Befehlstaktik* or "leading by orders." The command implications of *Auftragstaktik* were immediately apparent: leaders should resist too greatly impinging on how subordinates accomplished assigned missions. Doing so denied a junior leader's exercising of their own discretion to act in a manner that best suited a situation, a situation the subordinate might well better grasp than more distant leaders. Thus, the concept underlying mission command, "the practise of assigning a subordinate commander a mission without specifying how it is to be achieved" was introduced to the modern military age.[2]

Australian Defence Force Special Operations Command organizations have demonstrated a bias towards mission command[3] throughout their history, a bias especially appropriate given the nature of special operations. Sensitive and high-risk missions are almost always led by special forces personnel who are often relatively junior in rank yet assume a significant responsibility for mission success. This chapter will examine Australian special forces approaches to mission command, to include their application in doctrine, organization, during employment in the field, and as a philosophy of command. It concludes with a discussion of what the future might hold for special forces application of mission command in a world that is becoming increasingly digitized and interconnected.[4]

AUSTRALIAN SPECIAL FORCES — A BRIEF HISTORY

Australian special forces were first established when the "Special Operations Australia" force was created as part of the joint allied forces in Southeast Asia during World War II.[5] In one of the country's first recorded special operations, Australian personnel conducted a successful raid against Japanese forces in Singapore using a fishing boat, two canoes, and several limpet mines. From this meagre investment, the Allies destroyed more than 40,000 tons of Japanese shipping.[6] Australia chose to maintain a special forces capability following the war, creating a commando regiment, the Special Air Service Regiment, and a special forces signals squadron. These units participated in several conflicts in the years immediately following World War II including the Malayan Emergency, Indonesian Confrontation, and Vietnam War. Australia raised a Headquarters Special Forces (to be later renamed Special Operations Command) in the wake of the Hilton Hotel terrorist bombing in Sydney in 1978, thereby creating a unified entity serving as part of the Australian Army in lieu of standalone units as had previously been the case. By 2003, Special Operations Command would also grow an additional commando regiment, specialist engineer regiment, and bespoke training academy as well as other ancillary and operational military capabilities.[7]

Since the late 1970s, Australia's special forces have maintained a role as the country's counterterrorism "force of last resort." In this regard, Australia's Special Operations Command (SOCOMD) provides a "no-fail" military resolution capability to address terrorist attacks and other contingencies potentially requiring the use of lethal force in situations when the threat exceeds police agency capabilities or the government identifies an incident as "sudden and extraordinary."[8] Special operations task groups have also been widely deployed during international commitments in the years following the September 11, 2001 terrorist attacks in the United States. Acknowledged operational commitments include Kuwait (Operation Pollard), Afghanistan (Operation Slipper), and Iraq (Operations Bastille, Catalyst, Falconer, and Okra). Today, SOCOMD oversees in excess of 2,000 personnel, seven units, and a deployable headquarters. It is commanded by Special Operations Commander-Australia, an army major general responsible for maintaining Australia's special operations capability in addition to acting as the principal operational commander and adviser to higher command and the government more broadly regarding special forces contributions and other national security matters.[9] While special forces remain distinctly within the Australian Army's command chain, they are supported by the Air Force and Navy.

THE UTILITY OF SPECIAL FORCES

Conventional (otherwise known as regular) forces provide governments with a capability that seeks to safeguard national security through the provision of primarily large-scale military capabilities. While these forces have utility in addressing other facets of national security operations (e.g., humanitarian disaster relief and peacekeeping), they are primarily designed for high intensity combat.

Special forces are designed to support conventional forces during all phases of combat but are also structured to accomplish other missions, to include sensitive operations in complex terrain with little notice. These missions are often described as *unconventional* or *irregular.* Special forces units thereby provide policy makers with options different from, as well as complementary to, those conducted by conventional forces. Operations undertaken by special forces are generally

conducted in hostile, denied, or politically sensitive environments to achieve military, diplomatic, informational and/or economic objectives employing military and non-military capabilities for which there is no broad conventional force requirement.[10]

MISSION COMMAND IN THE AUSTRALIAN SPECIAL FORCES

Mission command in Australian special forces is essentially a framework that encourages innovation, imagination, and initiative within the bounds of strategic context and intent. What is unique about mission command in special forces is the ability for junior personnel to recognize context early through training reinforced by a rigorous personnel selection process and continuing professional development. The binding element that underpins much of this is the trust that exists amongst commanders and their men who have repeatedly deployed during the recent period of persistent conflict. Australian special forces have been continuously deployed domestically and overseas on operations throughout 2001 to the present.

Special forces leaders' application of mission command will always be conditional. In special forces, the "why" can matter more than the "what."

Special forces' incorporation of mission command as a command philosophy begins with selection programs. These programs seek to identify personnel who exhibit the traits necessary not just for command but also for leadership capable of independent decision making, grasping context when confronting highly complex problems, accomplishing time critical tasks, and operating in information-constrained environments. Amongst many of the traits sought are adaptability, physical and mental toughness, trainability, maturity, self-discipline, and a demonstrated ability to work well in a dynamic team environment.

Special forces mission command requires leaders with strong value systems who are able to build strong, trusting relationships with those around them as well as enable a safe risk management methodology to be adopted in training as a protective action to avoid failure on operations. The role that ethics play in military decision making and the execution of orders is more significant today than ever before. The asymmetric and amorphous threats that insurgency and terrorism present today have forced special forces to take a different view of their mission in terms of how they deal with threats. Operations into "ungoverned spaces" can no longer focus only on the military destruction of the enemy; they must also potentially win the "hearts and minds" of the host nation government and local population so as to erode the enemy's support structures and reduce the strength of their narrative. During Australian special forces operations in Afghanistan in 2010, a joint military-police operation into Gizab led to the popular overthrow of the local Taliban network and the reintroduction of legitimate Afghan government institutions. This operation demonstrated an understanding on the part of the special operations task group that the Afghan government (supported by coalition forces) had to generate an alternate narrative superior to that of the local enemy if there was to be enduring operational success. It did so via a series of shuras, veterinary and medical clinics, provision of safe passage for Afghan government officials, and guaranteed delivery of salaries to local police and military personnel. Operating within the higher commander's intent and in the absence of specific direction, these forces deposed the local Taliban network through its asymmetric delivery of essential services rather than direct action operations against enemy leaders favored at the time.[11]

Generally speaking, Australian special forces select their commanders at an early point in their careers, meaning they often have comparatively little leadership experience. Many of the assessments employed during selection test an individual's character and value system. This selection method contrasts sharply with most others, to include those in the commercial sector where leaders are generally identified through previous technical and functional performance.

Australian Special Operations Command emphasizes the flattening of organizational hierarchies. It is therefore not uncommon to see relatively junior personnel advising quite senior officers or officials.

A recent example of the importance of effective special operations junior leadership occurred during 2012 action in Helmand Province, Afghanistan. During an operation aimed at disrupting Taliban logistical systems, a single helicopter crashed on approach to its target, killing two Australian commandos and injuring several others. Despite the loss of almost one-third of the assault force, the 26-year-old sergeant on-scene commander effectively triaged the scene, organized the evacuation of his force's killed and wounded, facilitated recovery of the aircraft, and continued with the mission. The resulting disruption of supply networks denied the local Taliban of its ability to target Afghan government officials and civilians for weeks after the operation.[12]

After selection and training, special forces personnel are posted to their respective regiments whereupon leadership expectations commence immediately. Unlike in most other units where responsibilities equate to higher rank and seniority, special forces devolve considerable authority to relatively junior levels. Counterterrorism operations provide a ready example.

Recovery of the merchant vessel *Tampa* in August 2001 provides an example of relatively junior special forces leaders being placed in situations demanding well-considered actions within a strategic commander's intent.[13] The government of Prime Minister John Howard refused permission for the Norwegian freighter *Tampa*, carrying 438 rescued refugees, to enter Australian waters. This triggered an Australian political controversy in the lead up to a federal election and a diplomatic dispute between Australia and Norway. In this instance, the Australian government responded to this crisis by deploying a special forces element led by a major to board the ship and prevent it from approaching any closer to Christmas Island.[14]

Mission command in Australian special forces recognizes the importance of accurately gauging individual capabilities. War is unpredictable. The acceptance of environmental volatility and the chaotic effect this has on planning has led to recognition in Australian special forces that they must adopt an implementation methodology directing individuals to make decisions and achieve effects within broader organizational (i.e., mission) goals. Individuals are to exercise discretion in analyzing issues within the bounds of senior leaders' intents, changing plans as they see necessary to accomplish identified objectives.

One of the most recent examples of junior commanders sensing opportunities while operating within their commander's intent was in Iraq in 2003. Australia agreed to a coalition request to expand its special forces' area of operations at the end of March that year. The new area of operations included the Al Asad Airbase, 200 km west of Baghdad, one of Iraq's largest air bases. An entire SAS squadron concentrated to capture it on April 11, 2003. Australian commandos and the Incident Response Regiment accompanied them. Over the next 36 hours, these units cleared the massive base of a large number of armed looters while Royal Australian Air Force F/A-18 fighter jets provided overhead cover. More than 5o MiG jets and 7.9 million kilograms of explosives were captured. The task group further demonstrated its flexibility by clearing and repairing the runways using captured Iraqi military equipment and other systems borrowed from locals.[15] Tactical commanders, without orders, went beyond specified mission tasks, expanding their remit to also secure key infrastructure surrounding the airbase in the hope of preserving it for future Iraqi military use.

CHARACTERISTICS OF MISSION COMMAND IN AUSTRALIAN SPECIAL FORCES

Mission command is enabled through the characteristics that essentially define Australian special operations. They are encapsulated within doctrine, notably the *Australian Developing Doctrine Publication 3.12- Special Operations (Provisional)*.[16] These characteristics are as listed below.[17] Each receives attention in turn:

- Mature, ready, and relevant
- Versatility throughout the spectrum of conflict
- Independent, or in support, but inherently joint
- Adaptable
- Acceptance of risk and dislocation
- Detailed planning
- Precise effects
- Range of signature profiles
- Direct and indirect approaches

Characteristic 1: Mature, Ready, and Relevant

We noted that Australian special forces selection process recruits, screens, and accepts only those personnel capable of operating in austere, complex and unpredictable environments. Within this context, the special forces soldier must also be "ready"—that is, able to deploy anywhere across the globe at very little notice. This culture of hyper-readiness gives Special Operations Command, government and military strategists immediate response options otherwise not available.

An example of special forces readiness and its ability to anticipate future operations without express direction from higher command authorities was demonstrated in 2001 when Australian special forces redeployed from operations in East Timor to operations in Afghanistan in less than one month. Most, if not all of the planning for the retrograde from East Timor and the reconnaissance, insertion, and commencement of operations in Afghanistan sat on the edge of, or was slightly 'in front' of policy (driven by an understanding of the political narrative). This is all the more significant as an achievement given that Afghanistan was not assessed as a likely operating area by strategic planners until after the attacks on the World Trade Center in New York.[18]

Characteristic 2: Versatility Throughout the Spectrum of Conflict

Special forces personnel must be able to operate across the spectrum of conflict, inclusive of Australian Defence Force operations as well as outside of the traditional military domain. This includes supporting whole-of-government efforts such as disaster relief, aid distribution, and support to diplomatic missions. Special Operations Command has a wide scope of operating profiles and unique abilities—this includes the requirement to operate throughout the land, maritime, air, space and cyber domains, as well as within and external to Australia's sovereign territories and overseas interests.

An example of the versatility of special forces includes the operations to contain the East Timor independence group

Fretilin in 1999.[19] Australian special forces, operating with Fretilin leadership, including their recently released leader, Xanana Gusmao, successfully cooperated with ex-militants outside of the capital, Dili, to ensure that arriving Australian conventional troops did not mistakenly identify them as militia. This was achieved through the development of a sophisticated and considered plan to place special forces as the intermediaries between all protagonists upon realization that conflict between these parties was a key vulnerability to the overall plan. With no formal direction, but with an understanding of the strategic priority to protect the mission, special forces personnel developed various outreach and containment strategies, further strengthened by the trust and influence that junior special forces personnel develop as a core skill. This greatly reduced the likelihood of fratricide.

Characteristic 3: Independent, or in Support, but Inherently Joint

Australian special forces must be capable of supporting military and civil operations involving tasks ranging from those individual to undertakings involving large maneuver forces. Mission definition is sometimes inexact owing to dynamic circumstances. Indeed, there may be no mission specified but rather just loosely worded guidance based on an incomplete understanding of the operational environment. Special forces leaders must be able to distil these messages, extracting their underlying intent and articulating that intent to subordinates. During operations in Iraq in 2003, the special operations task group was tasked to clear the Khubaysa cement factory as part of their clearance operations of the Al Asad airbase. Sensing the value of the factory for the purposes of civilian reconstruction in post-conflict Iraq, the task group planned a purely non-kinetic clearance operation using low level fighter jet passes and 'cordon and call out' tactics rather than an overt assault which would have likely resulted in damage to the facility. The operation proved highly successful, with the local security forces surrendering and the capture of the cement factory complete without a single shot fired.[20]

Being able to simultaneously determine the objectives of military, civil government, coalition, and civilian organizations and understand how special forces can best address their common ends is fundamental to mission success. This is particularly true in an era of persistent conflict where special forces must provide non-traditional capabilities.

Characteristic 4: Adaptable

Adaptability implies that special forces personnel are capable of understanding the changing nature of the operating environment and can respond accordingly. Adaptation is temporal and relative to the threat; an effective rate of adaptation is one that outpaces the adversary and accounts for other environmental changes. The decentralization inherent in successful mission command enables a tempo that allows the maintenance of the initiative, essential for dictating successful mission outcomes.

Characteristic 5: Acceptance of risk and dislocation

Trusting that subordinates will effectively measure risk, understand risk and implement controls to mitigate it is an essential element of effective special forces mission command. Special operations are frequently conducted beyond the range of conventional forces support. Risk therefore has the dual components of risks to the mission and risks to the force. Gauging whether risk is acceptable or otherwise is possible only with complete understanding of senior commanders' intents that clearly articulate the relative importance of these two risk types and their relationship to sought-after strategic ends.

Characteristic 6: Detailed Planning

Decentralization within the scope of intent does not preclude detailed planning. On the contrary, the relative isolation of special forces units during many missions demands a readiness for multiple contingencies so that adaptation in the field does not disrupt mission success. Senior commanders will approve execution only if they are convinced there is a satisfactory union of understanding of both political and military intent, grasp of the operational environment, and acceptable assessment of risk.

Characteristic 7: Precise Effects

Special forces provide technological and human capabilities that are able to be prescriptive in nature and therefore precise in planning and effect. This degree of certainty allows strategic decision makers to better understand the character of the decisions that they are being asked to make, which, given their often strategic nature, have both a political and military dynamic. Given the potential consequences of mission failure, being able to define missions in terms of their intended effects is critical in gaining the trust and confidence of decision makers across all levels of command.

Characteristic 8: Range of Signature Profiles

Australia's special forces provide a range of response options in support of full-spectrum operations. These options and the various signature options available provide decision makers with the ability to vary the scale, scope, method and effect to suit the overarching commander's intent. This is critical for employing mission command at the operational level of war. Special forces should not be limited in insertion and maneuver options to a single platform only. They need a variety of military (and potentially non-military) platforms with which to give commanders a variety of mission profile options. These capabilities provide a full range of special operations options that allow for the development of missions that are manageable in terms of signature, risk, and supervision.

Characteristic 9: Direct and Indirect Approaches[21]

Owing to the nature of special forces, a full range of direct and indirect approaches are available to decision makers given the depth of capability resident within Special Operations Command. This allows special forces commanders to better develop multiple options for managing the uncertainty, fog, and friction of war. Given the nature of special operations, which can operate inside and outside of declared military activities, an ability for commanders to be able to achieve an "early effect" prior to the arrival of the remainder of the joint military force is critical to being able to generate superior tempo against the adversary, particularly in the early phases of conflict.[22]

CASE STUDY ON AUSTRALIAN SPECIAL FORCES' APPLICATION OF MISSION COMMAND: AFGHANISTAN 2006 — THE HUNT FOR OBJECTIVE NILE[23]

During combat operations in 2006 Afghanistan, an Australian commando platoon took part in a mission to neutralize a high value individual in support of Canadian and Afghan special forces.[24] The Australian special forces personnel would fight a series of ferocious running battles against a numerically superior and aggressive Taliban enemy led by the target, a senior level commander known as 'Objective Nile'. The success of the Australians involved provides an excellent example of effective mission command by those in the special operations community.

Background

In early July 2006, Taliban commander Objective Nile appeared with medium-high value on the coalition targeting list. Intelligence reported he was located in a compound complex in the village of Dehjawz-e Hasenzay approximately ten kilometers to the north of Tarin Kot, capital of Uruzgan Province in central Afghanistan. (See Map 10-2, page 289.) The report assessed there would be a two to three-man personal security detachment guarding Nile with a further two to three fighters in the local area capable of responding within thirty minutes. Planning commenced on a joint coalition operation to conduct a mission to neutralize Nile based on this estimate.

Force Composition

The Objective Nile assault force consisted of a troop of Canadian special forces (CANSOF) as the on-target direct action force augmented by a platoon of Afghan forces to serve as cordon security. The Australian commando platoon, designated by the call sign Bushranger Four Zero (B40), was the Quick Reaction Force (QRF). Intelligence, surveillance, and reconnaissance (ISR) would be provided by a MQ-1 Predator. A US Air Force AC-130 Spectre gun ship (BURNER 21) was to deliver any needed aerial fire support.

Concept of Operations

The CANSOF assault force would conduct a helicopter assault using two Australian CH-47 Chinook helicopters out of Kandahar Air Field (KAF) after re-fuelling at a Forward Air Refueling Point (FARP) at Tarin Kot. The QRF (B40) was to be vehicle mounted in three Bushmasters (BM) and four Special Reconnaissance Vehicles (SRV). B40 would conduct a preliminary move, posturing themselves north of forward operating base (FOB) Camp Davis in Tarin Kot. B40 was to commence their move at H-hour to a holding point south of the target compound and wait to be called forward on completion of the assault. The Australians would thereby be close enough to respond within ten minutes to any call for assistance but a sufficient distance away to not compromise the target. B40 was later to move forward to the vicinity of the target and secure a nearby landing zone (LZ) for the assault force's Chinook extraction, thereafter returning the FOB in Tarin Kot.

Synopsis of the Battle

Move to the holding point. At H-hour, B40 commenced its move to the holding point two kilometers south of the target. The skyline was alight with a series of flashes as BURNER 21 engaged pre-designated enemy locations with 40mm and 105mm cannon fire. Reports soon informed B40's commander that the assault force had landed on a hot LZ and was under heavy Taliban fire from a larger force than estimated. Two minutes later he received an order to assist the combined Canadian-Afghan force by occupying a blocking position close to the target. B40 was to relieve enemy pressure on the assault force, provide a potential fallback position for that force, and secure the extraction LZ.

Action on target. Aircraft touching down, the assault force came under heavy small arms, medium machinegun, and rocket-propelled grenade (RPG) fire resulting in one soldier killed in action and another two wounded. The Canadians and Afghans had broken into and cleared the target compound, killing several Taliban including Objective Nile during close combat engagements. Now receiving heavy fire from outside the compound, the assault force established a defensive perimeter. Pinned down by the intensity of fire and unable to maneuver, they found themselves encircled by a large enemy force.

Move to the blocking position. B40 came under ineffective small arms and rocket fire as they moved forward from the holding point to their blocking positions. Incoming fire intensified as they continued to move, B40's lead vehicle commander ordering his crew to attack the primary enemy position, suppressing the heavy weapons located there. B40's joint tactical air controller called in fire support from BURNER 21, destroying additional Taliban positions. B40 then continued forward to occupy the designated blocking position, continuing to contact the enemy en route.

The blocking position. On arriving at the blocking position, B40 was immediately engaged by small arms from a tree line 100-200m to their front. Additional fire from a nearby compound quickly joined the fight with machinegun and rocket fire that engaged both B40 flanks. BURNER 21 identified multiple enemy groups of 10 to 20 individuals around the Australian position. B40 defended the blocking position for the next hour despite being engaged from all directions and undergoing several Taliban attacks seeking to overrun the unit.

Extraction of the assault force. B40's commander understood the priority was evacuation of the assault force. He therefore broke through the Taliban perimeter and moved southeast to establish an alternative LZ. Supported by BURNER 21, the Canadians and Afghans withdrew from the compound, linking up with the Australians at the LZ while the Spectre gunship continued to engage Taliban forces swarming over the previously occupied compound, killing many and detonating an ammunition cache. BURNER 21 received a 30 minute on-station extension despite the aircraft running low on fuel as Chinooks moved forward from Tarin Kot to extract the assault force. The helicopters immediately came under heavy rocket and machinegun fire on landing. B40's commander ordered his vehicles to establish an attack by fire line, bringing massed firepower to bear until the completion of the extraction.

"Circling the wagons." B40 then turned its focus to its own extraction. The enemy would continue to engage the commandos from a series of ambush positions along its withdrawal routes. As they moved through built up areas south of the compound, lead elements came under concentrated fire at ranges as close as 30 meters. The intensity of incoming rounds forced B40 to form an all-round defense in order to avoid being overrun for the third time. BURNER 21 having departed to refuel, the platoon was without external support. It would be a half-hour before, a US Air Force B-1B bomber, call sign "CROW," arrived overhead roughly coincident with the return of BURNER 21. Both reported sighting of up to 120 Taliban forming for another attack against the Australians.

The break out. With the first rays of dawn appearing on the horizon, arrival of air support, and advantage offered by B40's advanced night vision equipment diminishing, the commando commander decided to break out for the return to Tarin Kot. Moving through the urban sprawl north of that urban area, B40's lead vehicles again came under close range fire by Taliban armed with additional heavy machine guns and rockets. Other enemy engaged vehicles farther to the rear from building rooftops, ranges sometimes being as close as three to five meters. Vulnerable if they remained in their vehicles, several B40 commandos dismounted and counterattacked. With no air support available in the densely packed urban terrain, B40 employed its full complement of weapons ranging from 84mm anti-armored fire to grenades and rifle fire as they sought out and destroyed the erstwhile attackers.

B40 was in a precarious position. Some commandos fired from vehicles while others cleared nearby buildings and streets. Maneuver was difficult given limited lines of sight, the ubiquitous presence of urban obstacles, and continued incoming enemy fire. Sensing the danger, a junior commander drove a heretofore unengaged Bushmaster into the ambush site as an armored shield to allow recovery of wounded Australian personnel. Crew members sustained fire as dismounted commandos returned to their vehicles, the convoy breaking out of the ambush to recommence movement to Tarin Kot.

Clear of enemy fire, B40 conducted a casualty assessment and ammunition redistribution. Spotting enemy personnel establishing mortar positions, the Australians engaged the targets as the platoon resumed movement. CROW, with BURNER 21 supporting calls for fire, dropped four five-hundred pound bombs on enemy redoubts outside the built-up area as B40 broke into the desert and returned to Tarin Kot.

Operational summary. Subsequent reporting from multiple sources confirmed that the Taliban were conducting a shura council (war meeting) in an adjacent compound with several key leaders from the Chora area during initiation of the Objective Nile mission. Those reports estimated that there were approximately 200 enemy in the village on the night of the operation.

The mission resulted in two commandos being lightly wounded by rocket fragments, this although several of their seven vehicles were riddled with bullet holes and one Bushmaster was immobilized by enemy fire. B40 returned to the field twenty-four hours later for Operation Perth, a ten-day clearance operation of a Taliban safe haven in the Chora valley.[25]

Special Forces Mission Command in Review

B40 showed great tenacity during four hours of near continuous fighting. In a series of highly kinetic, close combat engagements, the employment of mission command from the task group commander (a lieutenant colonel) down to individual commandos provided the difference between mission success and catastrophic failure. The following elements of mission command as applied by Australian special forces are especially worthy of note:

- **Decentralized decision making**: Key departures from the plan included the commando platoon commander's decision to move onto the target as well as others made when the platoon was confronted with large enemy forces during movement back to Tarin Kot.
- **An acknowledgement of the enduring frictions of war:** The platoon commander and subordinates adapted to situations in an information poor and highly dynamic environment. The case study demonstrated that there remained an ability for commanders to make clear, concise decisions within the bounds of senior commanders' intentions even during periods of complete chaos. Innovation, adaption, and audacity drove decision-making in the absence of direct supervision by those higher in the chain of command.
- **The investment of subordinates with trust, freedom of action, and the ability to effectively manage risk**: The B40 command team was empowered to make their own decisions determining their tactical fate throughout the mission. Senior commanders facilitated by providing resources when possible (e.g., coordinating support of B-1B bomber).
- **An emphasis on initiative and decisive action**: Bold, aggressive action taken by the platoon leader to support besieged Canadian and Afghan combined force after the neutralization of Objective Nile was key to the assault force's extraction.
- **Expertise**: Effective fusion of ISR, airborne fires, and ground maneuver at night in a coalition environment validated senior commanders' evaluations of junior leader expertise relevant to the mission.

- **Effective Joint Military Appreciation Process** application included designing plans incorporating the flexibility essential to successful application of mission command. The criticality of this flexibility in decision-making was apparent in the platoon leader's adaptations when needing to clear the target compound, more overtly assist extraction of the CANSOF-Afghan assault force, and respond to enemy resistance well beyond that expected.

CONCLUSION: THE FUTURE OF MISSION COMMAND IN AUSTRALIAN SPECIAL FORCES

Recent operational experiences have forced the Australian Defence Force to review the character of conflict and the role that the military plays. Greater emphasis is being placed on non-traditional operations and a new theoretical framework is beginning to emerge which is wider in breadth and multiagency in nature. This is known as "the comprehensive approach". It is for this reason that special forces have in many respects become an essential component of all military contributions in the post-September 11, 2001 era.[26]

Australian special forces personnel are specially selected, trained, and equipped; a critical factor given these operations often directly impact operational and strategic objectives. Special forces personnel conduct operations requiring discriminate and precise uses of force in a range of environments that at times are beyond the capability of other Australian Defence Force elements. Special forces provide Australia with a short notice force capable of conducting dynamic and highly versatile domestic and offshore operations. These operations can be unilateral, combined, or joint. They operate as an offensive force element at times independent of a support infrastructure using overt or clandestine techniques. To achieve this, special forces specialize in maintaining a technical and qualitative edge. These can include direct action tasks involving advanced infantry techniques as well as special operations-specific tactics, techniques, and procedures. Special forces leaders must therefore be proactive, free thinking, and have an attitude of continuous improvement in order to face the challenges of the future. Their soldiers employ a variety of specialist weapons and equipment not standard for conventional forces, requiring them to maintain proficiency in a multitude of specialized and often unorthodox combat skills. They deploy by conventional and unconventional means utilizing a range of joint and combined assets.

To meet these operational demands, special forces mission command places a distinct emphasis on harnessing potential advantages from its human and technological elements. Mission command provides a foundation for success when applied to collaborative planning to attain a tempo superior to that of adversaries and an ability to change tactics rapidly when necessary. These "mission command effects"—whether applied in the service of strategic, operational, or tactical objectives—seek to achieve overmatch by ensuring units are at the right place and time to achieve a decisive outcome. Through special forces' access to advanced command and control systems; technological reach-back; and ready access to intelligence, imagery, and joint fires, leaders applying mission command seek to accomplish their ends without the need for force-on-force attrition warfare.

The battlespace is changing. A networked Australian special forces imbued with a mission command ethos and a leadership biased towards action is potentially at risk against a force adept at cyber, information operations, deception, and unconventional operations. This constitutes both a risk and an opportunity for special forces. For special forces to retain its asymmetric advantage, it must look to develop countermeasures against these emergent capabilities in order to maintain overmatch regardless of the adversary. Special forces must also seek to embrace the opportunity that new technologies potentially offer. When combined with effective mission command, they represent a truly transformational combination that not only complements other elements of national power but additionally offers military commanders and the government a defeat mechanism in its own right.

ENDNOTES

1   Eitan Shemir, *Transforming Command* (Stanford: Stanford University Press, 2011), 29-53.

2   Land Warfare Doctrine 1, *The Fundamentals of Land Warfare* (Canberra: Australian Army, 2014), 45.

3   Mission tactics involve an orders and action-focused process providing subordinates with clear and concise descriptions along with appropriate resources. They are then held accountable for accomplishing assigned tasks. Importantly for all participants, context and a deep understanding of the relationship between the tactical operation and campaign objectives is essential to ensure the commander's intent is met without the need for scripted and overly process-driven orders (author's definition).

4   For the purposes of this essay, the term "special forces" denotes all Australian personnel serving in Australia's Special Operations Command. The author is aware of the alternative definitions employed in the United States military, including the use of SF and SOF as referring to quite distinct forces within US SOCOM.

5   Special Operations Australia (also known as the Services Reconnaissance Department), was an Australian military intelligence and special reconnaissance unit formed in April 1942, following the outbreak of war with Japan. See Alan Powell, *War by Stealth: Australians and the Allied Intelligence Bureau 1942–1945*, Carlton South, Victoria: Melbourne University Press, 1996.

6   Brad Manera, "Operation Jaywick," *Wartime: official magazine of the Australian War Memorial* 23 (2003): 53.

7   The incident occurred on February 13, 1978 when a bomb exploded outside the Hilton Hotel in Sydney. Three people were killed and eleven injured. The hotel was hosting the first Commonwealth Heads of Government Regional Meeting at the time. The Australian government deployed the Australian Army to the scene to provide security for the remainder of the meeting. See Tom Molomby, *Spies, Bombs and the Path of Bliss*, Sydney: Potoroo Press, 1986.

8   A "sudden and extraordinary" incident requires an immediate resolution to prevent further loss of life or strategic failure. Australian government, "National Counter Terrorism Plan," 2012, http://www.nationalsecurity.gov.au/Media-and-publications/Publications/Documents/national-counter-terrorism-plan-2012.pdf (accessed September 24, 2015).

9   Andrew Davies, *A Versatile Force: the future of Australia's Special Operations Capability* (Canberra: Australian Strategic Policy Institute, 2014), 9-14.

10   US Department of Defense, *Special Operations Force Posture Statement 2003/2004*, defenselink.mil/policy/SOLIC/2003_2004_SOF_posture_statement.pdf (accessed September 10, 2015).

11   Australian Department of Defence, "Gizab's Taliban commander captured," http://www.defence.gov.au/defencenews/stories/2010/May/0507.htm (accessed November 1, 2015).

12   Australian Broadcasting Corporation, "As it happened: five Aussie soldiers killed in Afghanistan," http://www.abc.net.au/news/2012-08-30/five-aussie-soldiers-killed-in-afghanistan/4233558 (accessed November 1, 2015).

13   "Australian National University Refugees between pasts and politics: sovereignty and memory in the Tampa crisis," http://

press.anu.edu.au/anzsog/immigration/mobile_devices/cho5so2.
html (accessed November 1, 2015).

14 Ibid.

15 SOCNET: The Special Operations Community Network,
ADF Special Forces Ops in Iraq 2003, http://www.socnet.com/
showthread.php?t=34645 (accessed November 17, 2015).

16 Australian Developing Doctrine 3.12- *Special Operations
(Provisional)* (Canberra: Australian Defence Force, 2011).

17 Author's interview with Commandant, Special Forces Training
Centre, September 9, 2015.

18 Special Broadcasting Service Australia *Timeline: Australian
Troops in Afghanistan since 2001,* http://www.sbs.com.au/news/
article/2013/03/26/timeline-australian-troops-afghanistan
(accessed November 1, 2015).

19 For more on Fretilin, see John Crawford and Glyn Harper,
*Operation East Timor: The New Zealand Defence Force in East
Timor 1999-2001*, Auckland: Reed, 2001, 11-13.

20 SOCNET: The Special Operations Community Network,
ADF Special Forces Ops in Iraq 2003, http://www.socnet.com/
showthread.php?t=34645 (accessed November 17, 2015).

21 United States Department of Defense, *Joint Operation
Planning*, http://www.dtic.mil/doctrine/new_pubs/jp5_0.pdf
(accessed November 1, 2015).

22 The "direct" approach is one that is best suited to supporting
joint force conventional operations and can be planned
and executed with little more than an understanding of
the operational level effects required. It is characterized by

technologically enabled, small unit precision lethality, focused intelligence, and interagency cooperation integrated into the joint digital battlefield. The "indirect" approach is somewhat more nuanced. It can support both military and nonmilitary operations while providing more discreet and disparate effects that support missions where the military signature may need to be varied. This approach seeks to empower relevant elements in partnership with special forces, both inside and outside the military domain. Examples include support to a whole-of-government effort to recover a hostage or equipment from a foreign government or terrorist group. In this instance, it is important that the special forces personnel involved have a sense of the full range of strategic sensitivities and issues surrounding the use of intermediaries or proxy forces when special forces access is denied.

23 The following case study is drawn from the author's interviews with elements of the special forces Task Group 637 Rotation III Commando Force Element, specifically WO1 M, MAJ J, and WO2 R as well as a key staff member from the task group headquarters (MAJ L). Actual names and tactical call signs have not been used to protect operational security.

24 "Why SGT Brett Wood was awarded a medal for gallantry," *Herald Sun*, http://www.heraldsun.com.au/news/why-sgt-brett-wood-was-awarded-a-medal-for-gallantry/story-e6frf7jo-1226061878021 (accessed November 1, 2015).

25 The toll inflicted on the enemy cannot be confirmed for several reasons, but it has been conservatively estimated from reliable sources to be in excess of seventy killed and many more wounded.

26 UK Government House of Commons Defence Committee "The comprehensive approach: the point of war is not just to win but to make a better peace," 2009-2010, 2011-2012. http://www.publications.parliament.uk/pa/cm200910/cmselect/cmdfence/224/224.pdf (accessed October 1, 2015).

12

MISSION COMMAND DURING THE QUEENSLAND NATIONAL EMERGENCY, 2010-11

Major General Chris Field, Australian Army

This chapter applies six mission command principles to actions of a seven-person Australian Defence Force (ADF) planning team working for Australian Army Major General (MAJGEN) Mick Slater during and immediately after the Queensland national emergency of 2010-11.[1] Under Slater's leadership, this team formed and then dispersed within twenty-one days. Working to MAJGEN Slater's intent, the team's service proved vital in creating a plan to transition Queensland from crisis to recovery. Through MAJGEN Slater's employment of the doctrine of mission command, he provided the team a framework for engagement with Queensland's civilian leadership and other supporting military organizations in accomplishing this end.

Leadership responding to the 2010-2011 Queensland national emergency was diverse, including as it did Commonwealth (federal), state, local government, businesses, trade associations, nongovernmental organizations, and community-based representatives. ADF planning support was led by MAJGEN Slater. Slater focused primarily on developing a campaign plan in partnership with the Queensland's government leadership to incorporate all relevant available resources. This whole-of-government effort created the means of bringing these diverse assets to bear in the service of Queensland's citizenry.

MAJGEN Slater is a 2004 graduate of the United States (US) Army War College. He had from that opportunity gained an understanding of the requirements for a "genuine whole-of-government approach to [Australia's] strategic level planning for operations" and the "need to expand [the ADF's] training at the operational level to include more cooperation from civilian organisations, especially non-government organisations."[2] Slater's war college experience also influenced his application of mission command. The ADF's mission command doctrine aligns closely with that of the US. For example, the table below reflects that Australian Army capstone doctrine, *The Fundamentals of Land Power*, has parallels to all six mission command principles espoused by the US Army. For simplicity, this chapter therefore applies the US Army's six mission command principles to the 2011 actions of the ADF planning team in support of Queensland's recovery.

Australian Army	US Army
Grant trust and freedom to subordinates	Build cohesive teams through mutual trust
Junior leaders possessing a detailed understanding not only of the immediate tactical commander's intent, but also of the broader operational and strategic situation	Create shared understanding
Develop a clear expression of the senior commander's intent	Provide a clear commander's intent
Subordinates are expected to apply individual judgement in achieving the commander's intent, regardless of changing situations	Exercise disciplined initiative
Assign a subordinate commander a mission without specifying how the mission is to be achieved	Use mission orders
Junior leaders are expected to seek opportunities to immediately pursue their commander's intent once tasked and resourced	Accept prudent risk

Table 12-1: Six Australian Army and United States Army Mission Command Principles[3]

SMALL TEAMS AND MISSION COMMAND

A small, broadly experienced and educated team is more agile than a larger team of specialists. Multiple specialists can insulate the larger teams of which they are a part by reinforcing self-belief that members can account for all possible solutions. A small well-led team is more likely to understand that it will on occasion need to seek advice.

Regarding large teams, Jim Storr observed that military headquarters

have become much bigger and tend to produce worse plans and take much longer to produce them…. Process has become an end in itself. [Paraphrasing Brook's Law,] there is an optimum size for groups of human beings who interact. It is a balance between dividing a job up between more people to reduce the time taken against the increased time needed to brief all the members of a larger group.[4]

Effective small planning teams iteratively frame and articulate fundamental ideas while seeking the counsel of leaders and consulting experts as necessary. Through the prism of mission command, this chapter explains how the ADF planning team sought counsel, consulted with outside experts, and delegated tasks as they collaboratively developed the plan that would drive the campaign known as Operation Queenslander.

THE QUEENSLAND NATIONAL EMERGENCY 2010-11

Queensland

Australia is the:

World's sixth largest country by area, 7.6 million square kilometres [making it] slightly smaller than the US contiguous 48 states, [and the] driest inhabited continent on earth, making it particularly vulnerable to the challenges of climate change [and natural disasters] such as cyclones along the coast, severe droughts, and forest fires.[5]

At 1.7 million square kilometers, Queensland comprises over twenty per cent of the Australian land mass and is the second largest Australian State. It is two and a half times the size of Texas. More than half of Queensland's population of four million live outside the capital city of Brisbane.[6] This is unique; the other five Australian states each include more than half of their populations inside their respective state capitals.

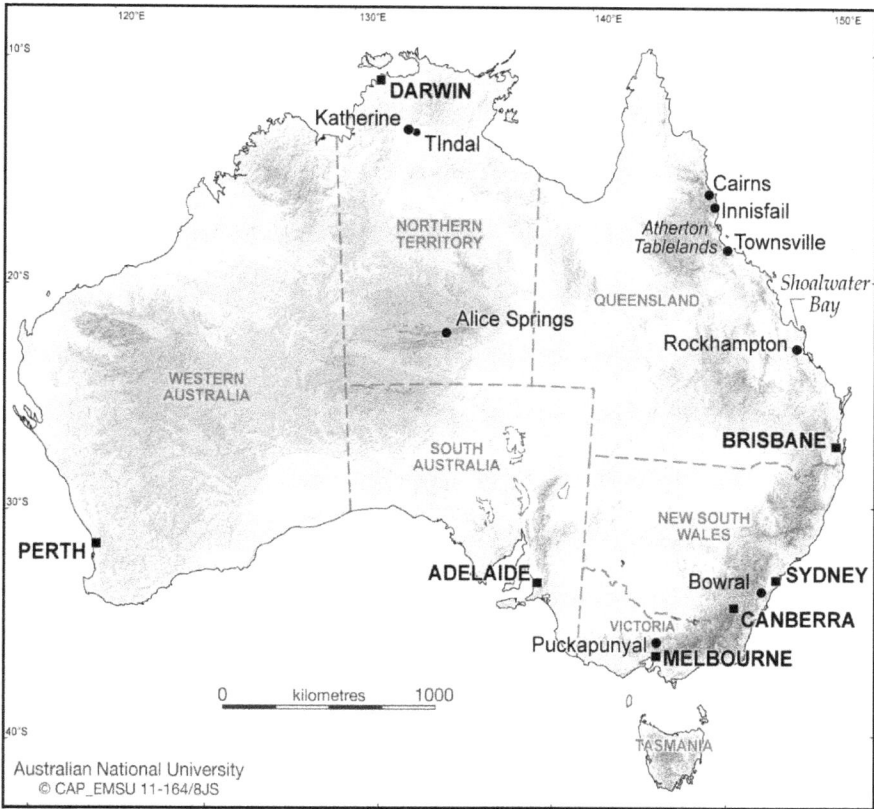

Map 12-1: Australia

The Queensland economy is diverse, including agriculture, natural resources, construction, tourism, manufacturing and service sectors. The state of Queensland is the world's largest seaborne exporter of metallurgical coal. Liquefied natural gas is also a major state export.[7]

Floods and Cyclones, 2010-11

On January 10, 2011, a flash flood struck Toowoomba and the Lockyer Valley west of Brisbane. Within days, the waters had flowed east to flood Ipswich and Brisbane. A total of thirty-five people lost their lives in the flooding, including twenty-one in Toowoomba and the Lockyer Valley. Widespread property damage accompanied the deaths.[8]

In February 2011, Tropical Cyclone Yasi crossed the far north Queensland coast near Mission Beach. The category five cyclone brought high winds, heavy rainfall, and five-meter storm tides. One person was killed. Significantly, Yasi was the third cyclone to impact north Queensland in the 2010-11 season. Tropical Cyclone Tasha (category one) had crossed the coast early on December 25, 2010 while Tropical Cyclone Anthony (category two) crossed near Bowen on January 30, 2011.[9]

The 2010-2011 Queensland national emergency encompassed damage wrought by both the above-noted floods and Cyclone Yasi. As stated earlier, flooding in Queensland began on November 30, 2010. However, the federal government did not formally declare a national emergency until December 21, 2010. The national emergency ended on February 14, 2011.[10] The combined disasters would eventually see over seventy-eight per cent of Queensland declared a disaster zone with two and a half million people suffering the flooding's effects.[11] Fifty-nine rivers flooded. Twelve broke flood records. Nineteen thousand kilometers of state and local roads and twenty-nine percent of Queensland's rail network were damaged.[12] All seventy-three Queensland local government jurisdictions were declared disaster areas.[13]

FIRST RESPONDERS

Government policy designates ADF responsibilities to counter both international and domestic threats to the security and safety of the Australian people.[14] Important as domestic contingencies are, however, policy as outlined in the *Defence White Paper 2009* explicitly states a that Australia's "most basic strategic interest remains the defence of Australia against direct armed attack."[15]

Yet the white paper also acknowledges a "vital role" for the ADF in "supporting domestic security and emergency response efforts" that include disaster relief.[16] Specific domestic threats recognized include "natural disasters such as cyclones, earthquakes, floods and bushfires [that] can also threaten the security and safety of the Australian people."[17] In meeting these responsibilities, the ADF's response to the 2010-2011 Queensland national emergency was rapid, extensive, and effective.

Australian Defence Force support to Queensland was not limited to senior level coordination and planning. Brigadier Paul McLachlan (and later as the emergency eased, Colonel Luke Foster) commanded Joint Task Force 637 (JTF 637) for Operation Queensland Flood Assist in the south of the state from December 2010 to January 2011. Nearly two thousand defense personnel deployed as part of the task force. Other Ministry of Defence support included the Her Majesty's Australian Ships (HMAS) *Huon*, *Paluma*, and *Shepparton*. The Royal Australian Air Force (RAAF) provided both C-130 Hercules and C-17 Globemaster aircraft.

Joint Task Force 637 troops assisted in transporting displaced persons, providing other logistical support, removing debris, and conducting searches for survivors.[18] Elements from four of the Australian Army's six regular army brigades and the army reserve's 11th Brigade supported response efforts. The regular army brigades represented included the 7th Brigade from Brisbane and three enabling brigades: the 6th Combat Support Brigade, 16th Aviation Brigade, and 17th Logistics Brigade.

Complementing the service of JTF 637, Brigadier Stuart Smith's JTF 664 supported Operation Yasi Assist in Queensland's north during February 2011. The task force included over 1,200 ADF personnel who brought their specialist aviation, engineering, health, and logistic skills to those in need. Assets included HMAS *Tarakan*, HMAS *Brunei*, RAAF aircraft, and the regular army's 3rd Brigade (from Townsville) and further elements from the three enabling brigades also lending support as part of JTF 637.[19] Both in Queensland's north and south, these JTFs performed a "vital role [in] supporting domestic security and emergency response efforts [including] disaster relief."[20]

OPERATION QUEENSLANDER — MISSION COMMAND SUPPORTING QUEENSLAND'S DISASTER RESPONSE, RECOVERY AND RECONSTRUCTION

Australia's Department of Defence seconded the the army's seven-person planning team to the Queensland state government during the January-February 2011 period to assist in reconnecting, rebuilding, and improving Queensland's communities and economy.[21] Their work was separate from but complementary to the efforts of JTF 637 and JTF 664 as the plan created by the team members would directly impact the nature of joint task force support provided. The result of the planners' work was Queensland's strategic reconstruction plan: *Operation Queenslander, The State Community, Economic and Environmental Recovery and Reconstruction Plan 2011–2013 (The State Plan).*[22] It was, in short, a campaign plan for Queensland's disaster recovery.[23]

In the words of General (British Army, retired) Sir John McColl, campaign planning

helps produce order out of chaos. [A campaign plan] forces you to write down what you want to happen. It imparts intent coherency. It is also a process which, if used properly, integrates lines of operation. It uses an internationally agreed and defined language so we all understand what we mean when we use terms such as centre of gravity, decisive points, and lines of operation. . . . A campaign plan is not a panacea, but it is a help. The danger with the process [of campaign planning] is that it can become a goal in itself; agreement is often hard fought and painful. Arriving at an agreed campaign plan can be exhausting, but of course the final plan is irrelevant unless it leads to delivery.[24]

The campaign to ensure Queensland's effective recovery was expanded following the release of *The State Plan* and departure of the ADF planning team from the newly created Queensland Reconstruction Authority. The strategic plan was provided with greater fidelity through supporting operational and tactical plans respectively titled *The Community, Economic and Environmental Recovery and Reconstruction Implementation Plan 2011-2013 (The Implementation Plan)* and *A Guide to Local Community, Economic and Environmental Recovery and Reconstruction Planning (The Local Plan)*.[25]

As noted in the introduction, this chapter analyzes the actions of the ADF planning team during and immediately after the Queensland national emergency of 2010-11 in light of six US mission command principles. Once again, the six principles are:

- Build cohesive teams through mutual trust
- Create shared understanding
- Provide a clear commander's intent
- Exercise disciplined initiative
- Use mission orders
- Accept prudent risk.[26]

This remainder of this chapter considers the influence the small ADF planning team had via application of mission command in accordance with each of these principles.

Build Cohesive Teams through Mutual Trust

Trust is gained or lost through everyday actions more than grand or occasional gestures. Trust is based on personal qualities such as professional competence, personal example, and integrity. [People] must see values in action before they become a basis for trust. Trust comes from successful shared experiences and training… [as well as] two-way communication and interaction between [people].[27]

US Army Doctrine Reference Publication 6-0, *Mission Command*

Most of the ADF planning team had never met prior to their arrival in Queensland. They were strangers thrust into crisis amidst unfamiliar state government leaders and other members of the civilian bureaucracy. Led by newly promoted army Lieutenant Colonel Jim Hammett, the team recognized they needed advice were they to be effective. The member represented all three military services and both regular and reserve personnel. While the team's composition was ad hoc, their common education in joint planning helped them quickly form a cohesive team formed on a common understanding of the approach to be taken.

Despite government policy requiring ADF preparedness to support domestic security and emergency response efforts, Australia's Department of Defence remains an internationally focused organization.[28] Formal doctrine for interoperability between the ADF and Australian state governments was limited in 2010 and remains so today. For example, detailed guidance on interoperability with Australia's states is absent in the *Defence White Paper 2009* despite its observation that "the ADF and other agencies of Defence have significant capabilities that can be used to support…emergency response and disaster recovery."[29]

In a mission command context of providing clear intent for an organization to implement an appropriate plan, *Defence White Paper 2009* recognizes the integration of organizations such as the

ADF planning team into national emergency response planning:

> In some limited cases, Defence capabilities will need to be designed for and dedicated to domestic security and emergency response tasks where they provide specialised capacities beyond the ability of other Australian government agencies and other Australian jurisdictions to efficiently develop and maintain.[30]

Three elements in particular provided development of the mutual trust fundamental to mission command during the interfaces between the Queensland government, federal government departments, and ADF planning team. First, the disaster response work conducted by JTFs 637 and 664 quickly demonstrated defense competence to the Queensland people and their government representatives.

Second, MAJGEN Mick Slater, who did not command the two JTFs—but had previously commanded the 1st Australian Division in Brisbane—was familiar to many in the state government. He was appointed to lead the development of a plan to guide Operation Queenslander on January 4, 2011. Slater quickly developed effective working relationships with the Queensland Premier, Ms. Anna Bligh, and Queensland Reconstruction Authority Chief Executive Officer, Mr. Graeme Newton. MAJGEN Slater's leadership and familiarity to these civil leaders further facilitated his planning team's rapidly gaining a trusted place within the hierarchy of six key Queensland government departments.[31]

Third, and resulting directly from the first two factors, the cooperation and support of Queensland's senior departmental leadership did much to expedite the trust bestowed in the planning team at lower civil echelons. Those manning positions where "the rubber hit the road" readily internalized the trust shown in the planning team by their seniors. That the senior leadership of these departments so quickly facilitated buy-in in turn reflected the importance of familiarity at the highest echelons of the campaign. That MAJGEN Slater was a familiar and trusted entity in the offices of Anna Blight and Graeme Newton argues for regularly linking civil leadership with those in the military likely to be tasked with support during a crisis.[32]

Create Shared Understanding

Shared understanding and purpose form the basis for unity of effort and trust. Commanders and staffs actively build and maintain shared understanding...by continual collaboration throughout the operations process (planning, preparation, execution, and assessment).[33]

US Army, Doctrine Reference Publication 6-0, *Mission Command*

General Slater's duties with the Flood Recovery Task Force and later as head of the Queensland Reconstruction Authority limited the time available for direct involvement with the planning team. These responsibilities meant he was frequently absent from Brisbane attending to community engagement and state development matters.

Slater drew heavily on his trust in the ADF planning team during these frequent absences. Despite not knowing each person in the planning team, he knew enough about the individuals to feel he could trust the entire team. Ultimately, Slater believed those manning the planning team had the skills and capacity to effectively integrate into Queensland state departments as necessary to create an effective *State Plan* for Operation Queenslander.

Slater's ability to transfer civil authorities' trust in him to those on his planning staff was invaluable, but such initially fragile acceptance would quickly dissipate were civil authorities not to understand how the seven ADF members could assist in Queensland's recovery. Short on time and needing to at once compile information essential to a coherent plan, process such information quickly and effectively, and prepare a plan both appropriate and acceptable to those responsible for its implementation, the planners turned to the ADF's proven Joint Military Appreciation Process (JMAP). Drawing on both information from their civil partners and an understanding of Slater's intent,[34] the seven-man team deliberately employed the existing Queensland government framework based on six lines of reconstruction:

- Human and Social
- Economic
- Environment
- Building Recovery
- Roads and Transport
- Community Liaison and Communication.

To enhance the development of a shared understanding between the military and civilian communities yet further, planners modified the Joint Military Appreciation Process by "de-militarizing" its terminology, further making the process comprehensible to those previously unfamiliar with it. For example, wording of intent statements in *The State Plan* was articulated in terms of "why, how, and outcome" instead of the traditionally military phrasing in terms of "purpose, method, and end state."[35] Such efforts reinforced civilian authorities' confidence in the military representatives' understanding of Queensland's needs no less than the efforts of the latter were more readily grasped by those in state and local government offices and others supporting recovery.

With the lines of operation framework in place, JMAP enabled the planning team to guide Queensland state authorities through the response, recovery and rebuilding process. Adapting the military planning process to civil needs meant each department was able to more rapidly visualize how their roles and responsibilities fit in the orchestrated whole that was the campaign guided by *The State Plan*.

Shared understanding requires collaboration. In Queensland, the primary integrators were the leaders of the six state government departments and subcommittees representing their half dozen lines of reconstruction. (The six reconstruction sub-committees included members from federal, and local governments, commercial businesses, peak industry organizations, nongovernment organizations, and community citizenry in addition to state representatives.) Many of these experienced people intuitively understood that the planning team was an ally that could assist them in building a plan to guide Queensland's recovery. In turn, these leaders recognized the essentiality of their expertise in facilitating the cooperative ultimate product. These leaders were accountable for the accuracy of information included in the plan, measuring the effects of their department's efforts, and—ultimately—the success or failure of Queensland's recovery.

Provide a Clear Commander's Intent

The commander's intent is a clear and concise expression of the purpose of the operation and the desired end state that…provides focus to the staff, and helps subordinate and supporting commanders act to achieve the commander's desired results without further orders, even when the operation does not unfold as planned.[36]
US Army, Doctrine Reference Publication 6-0, *Mission Command*

Unlike the traditional ADF command and control structures found in JTFs 637 and 664, there was no "commander" for Operation Queenslander. There was no one person dispensing a perfectly formed commander's intent to lead Queensland's recovery. Instead, in accordance with the protocols of democratic government, the Queensland premier, Ms. Anna Bligh led the state with her ministers and senior departmental leaders. Thousands of community leaders in the public and private sectors simultaneously worked to repair the damaged state.

Armed with MAJGEN Slater's intent regarding activities *internal* to their team, the ADF planners sought to identify and articulate the largely unstated "commander's intent" representing the aspirations of Queensland state leaders. Successful interface with those individuals relied on supporting existing structures, maintaining the trust of those responsible for implementing the campaign plan, and understanding the immediate and longer-term vision Queensland's leadership had during and after recovery. Mission command for the ADF planners therefore relied on practicing what Nelson Mandela referred to as "leading from behind":

It is better to lead from behind and to put others in front, especially when you celebrate victory when nice things occur. You take the front line when there is danger. Then people will appreciate your leadership.[37] . . . A leader . . . is like a shepherd. He stays behind the flock, letting the most nimble go out ahead, whereupon the others follow, not realizing that all along they are being directed from behind.[38]

Here again, ADF planning team members adapted themselves to the structures of the supported Queensland civilian government rather than trying to rigidly retain traditional military command processes and relationships. The ADF planning team therefore overcame sometimes blurred or overlapping leadership boundaries by continually liaising with Queensland's leadership in pursuit of developing a civil-military intent suitable for guiding actions of all participants at every echelon. Mission command in the context of the ADF planners' contributions to Operation QUEENSLANDER is therefore more a model for enabling campaigns outside a traditional military environment than an exemplar in the traditional sense. The ADF planning team incorporated familiar elements of mission command—trust, expertise, reliability, and ever-improving familiarity—to better understand Queensland leaders' recovery ambitions and, thereby, develop an intent applicable across the many varied tasks necessary to the state's recovery.

Communicating a clear commander's intent is a challenge no less during peacetime contingencies than war. In war it is the enemy that seeks to resist and disrupt the best of plans. During Operation Queensland, it was floods and cyclones that influenced and sometimes disrupted even the best of plans. Ultimately, planners must heed General Eisenhower's dictum that

> Plans are worthless, but planning is everything. There is a very great distinction because when you are planning for an emergency you must start with this one thing: the very definition of "emergency" is that it is unexpected; therefore it is not going to happen the way you are planning.[39]

Its scope of tasks meant Queensland's national emergency required mission command no less than did General Eisenhower's operations during World War II. State leadership could no more come up with a plan sufficiently detailed to cover every contingency than could that general's staff. Allied forces awoke to new challenges each day in Northern Africa or Western Europe; conditions in Queensland continuously evolved as nature devastated infrastructure and threatened lives. Because a neat, well-crafted commander's intent was not readily at hand, ADF planners defined and tested relevant guidance from civilian leaders for the Queensland national emergency problem to determine their intent. This defining and testing was done in conjunction with Queensland leaders until agreement was achieved.

Employing this iterative approach to understanding the intent of Operation Queenslander, the ADF planners eventually defined five primary campaign objectives:

1. Maintain Queensland's self-confidence through continuing support and restoration of essential services to affected communities.

2. Implement a comprehensive and integrated state community, economic, and environmental recovery and reconstruction plan (*The State Plan*) to restore community structures, public infrastructure, support economic growth, and facilitate environmental rehabilitation.

3. Maintain engagement with affected communities, local government, and industry groups during the reconstruction effort.

4. Enhance the resilience of Queensland and Queenslanders, informed by the recommendations from the Commission of Inquiry into the 2010/2011 flood events.

5. Continue implementation of *Toward Q2: Tomorrow's Queensland*.[40]

The fifth was the objective most amenable to treatment as a "commander's" intent for Operation Queenslander—continue implementation of *Toward Q2: Tomorrow's Queensland*— for it was *Toward Q2* that defined the campaign's end-state.[41] Operation Queenslander sought momentum from *Toward Q2: Toward Q2* provided the vision for taking the state beyond the national emergency. Ultimately, *Toward Q2* sought to not only to describe Queensland after recovery, but to establish understanding of how the recovery was a step toward a Queensland even better than pre-disaster.

Exercise Disciplined Initiative

Disciplined initiative is action in the absence of orders, when existing orders no longer fit the situation, or when unforeseen opportunities or threats arise. Commanders rely on subordinates to act. A subordinate's disciplined initiative may be the starting point for seizing the tactical initiative. This willingness to act helps develop and maintain operational initiative . . . to set or dictate the terms of action throughout an operation. [42]

US Army, Doctrine Reference Publication 6-0, *Mission Command*

Inserted amongst Queensland government departments, and left largely unsupervised, the seven-person ADF planning team maintained a bias toward taking action. They maintained this bias even as the Queensland national emergency evolved in early February 2011 from flooding in Southern Queensland to cyclones in Northern Queensland.

As noted earlier, the ADF planning team was assisted in exercising disciplined initiative through cooperative support from the six lead Queensland government departments. The departmental leadership thoroughly welcomed and engaged the planners. The team consistently gained access to senior leaders throughout the Queensland government via this cooperation.

However, leadership in Queensland's communities was the decisive factor enabling disciplined initiative. As earlier stated, more than half of Queensland's population of four million live outside the capital city of Brisbane, a situation unique in Australia. [43] Importantly, the dispersed nature of the Queensland population, among other issues, enabled the planning team to witness and encourage community-based disciplined initiative during the emergency.

A decentralized population spreads leaders over a wider area. In a crisis, decentralized leaders steady, reassure, and mobilize communities. Locally-tailored decisive action by decentralized leaders imbues community confidence and enhances resilience. Decentralized leadership means planners and well-led communities rapidly build mutual trust, create shared understandings, and exercise disciplined initiative. During the 2010-11 Queensland national emergency, decentralized leadership enabled by the application of mission command helped rapidly identify and complete essential tasks.

Working with its decentralized population, the Queensland state government departments were well positioned to monitor natural disaster conditions given their connections with local government authorities, nongovernmental organizations, and communities. The efforts of all six lines of reconstruction well orchestrated and shared metrics used in measuring existing conditions and progress.

Working with decentralized Queensland state government and community leadership, the ADF planning team formalized guidance to enable reconstruction actions. The ADF planning team planned, collaborated and wrote the Operation Queenslander Implementation Plan, the State's operational level plan, which detailed tasks and metrics common across the state. Procedures outlined in the Implementation Plan enabled the Queensland Reconstruction Authority to provide comprehensive monthly board reporting to the federal government in Canberra, state authorities, and local stakeholders.[44]

In summary, disciplined initiative following the Queensland national emergency arose through decentralized, diverse and well-led communities. These communities, supported by the ADF planning team and Queensland government, demonstrated a willingness toward action in assisting Queensland's recovery. Queensland communities seized the initiative in rebuilding their own State.

Use Mission Orders

An order should not trespass upon the province of a subordinate. It should contain everything that the subordinate must know to carry out his mission, but nothing more…. Above all, it must be adapted to the circumstances under which it will be received and executed.[45]

US Army, Field Manual 100-5,
Tentative Field Service Regulations—Operations Oct 1939

Commanders use mission orders to assign tasks, allocate resources, and issue broad guidance. Mission orders are directives that emphasize to subordinates the results to be attained, not how they are to achieve them.[46] They provide subordinates the maximum freedom of action in determining how to best accomplish missions and seek to maximize individual initiative while relying on lateral coordination between units and vertical coordination up and down the chain of command.[47]

The Queensland nation emergency was chaotic. The challenges continuously evolved from its start date in December 2010 into February 2011. Mission orders ideally suited the seven-person ADF planning team in such an environment. As noted earlier, the many responsibilities allocated to MAJGEN Slater necessitated he be able to trust the members of his planning team. This was all the more important given the dynamic nature of the environment and constant requirement to interface with and apply guidance received from the reconstruction sub-committees representing Operation Queensland's six lines of operation.

Carl von Clausewitz might well have been describing a commander contemplating the approach to mission command as he wrote of the strength and courage demanded to parse clarity in times of chaos:

If the mind is to emerge unscathed from this relentless struggle with the unforeseen, two qualities are indispensable: first, an intellect that, even in the darkest hour, retains some glimmerings of the inner light which leads to the truth; and second, the courage to follow that light wherever it may lead.[48]

Not only must the commander have the strength of mind to avoid being overwhelmed, so too must the senior leader have courage both in the sense of his/her own willingness to maintain a steady way ahead and in letting those at lower levels determine just what path that light illuminates given a clear intent from above.

The joint military appreciation process (JMAP) assisted the ADF planning team in its development of mission orders as they sought to chart that course. JMAP provided a proven approach for digesting large amounts of information and thereafter iteratively framing the material to abet understanding and defining solutions in the service of Queensland recovering from the national emergency. This process enabled both the planners and those serving on the six reconstruction sub-committees the time to refine their own thinking. Quite quickly, leaders throughout the state began to envision the way ahead toward a Queensland that had moved beyond the disaster. The civil-military cooperation inherent in applying the joint military appreciation process gradually brought order to chaos. So too, the capabilities provided by the ADF planning team gave Queensland leadership space to think. Without them, Operation Queenslander would have been delayed as leaders struggled with tasks crucial to reconnecting with the electorate and rebuilding devastated Queensland.

Mission orders worked because the ADF planning team knew what their commander intended and with whom they needed to coordinate to ensure the plan had "buy in" once it was disseminated. Though they did not know when the plan was required, such coordination ensured all relevant parties could conduct internal planning and preparations, thereby fast-tracking their efforts to minimize delays once the plan was releases. The freedom of action granted by MAJGEN Slater was fundamental to the responsiveness necessary to meet both the requirements inherent in addressing the six lines of operation and and constantly evolving conditions throughout the stricken area.

Accept Prudent Risk

Commanders accept prudent risk when making decisions because uncertainty exists in all military operations. Prudent risk is a deliberate exposure to potential injury or loss when the commander judges the outcome in terms of mission accomplishment as worth the cost. Opportunities come with risks. The willingness to accept prudent risk is often the key to exposing enemy weaknesses. [49]
US Army, Doctrine Reference Publication 6-0, *Mission Command*

The greatest risk for Operation Queenslander was ensuring value for money on Australian federal investments provided via Natural Disaster Relief and Recovery Arrangements (NDRRA). [50] NDRRA are a joint funding initiative of the Commonwealth and state governments to provide disaster relief and recovery payments and infrastructure restoration funding. Most relief measures under NDRRA were funded seven-five per cent by the Commonwealth and twenty-five per cent by Queensland. [51] Eventually, Operation Queenslander enabled the effective expenditure of Aus$14 billion in NDRRA funds for Queensland. [52]

To enable prudent risk, those executing Operation Queenslander incorporated a value for money framework into NDRRA analyses. This was achieved via a three-tiered review process consisting of: [53]

1. Tier One: Desktop review
2. Tier Two: Secondary review
3. Tier Three: On-site review[54]

A project whose status was unclear or thought unlikely to achieve value for money was transitioned to Tier Two. Some projects progressed to Tier Two and Tier Three even where they have been assessed as value for money at Tier One. Where a project was assessed as not achieving value for money, the process agreed with Emergency Management Australia (EMA) for non-value for money projects was triggered.[55]

Accepting reasonable risk is not gambling. Gambling, in contrast to prudent risk-taking, is staking the success of an entire action on a single event without considering the hazards should the event not unfold as envisioned.[56] With Aus$14 billion in NDRRA for Queensland, value for money remained a consistent risk for the State. Following the chaos of the Queensland national emergency, the carefully designed and agreed upon value for money framework reduced the risk to taxpayer money from gamble to prudent risk.

CONCLUSION

This chapter applies six mission command principles to the actions of a seven-person ADF planning team's actions undertaken during and immediately after the Queensland national emergency of 2010-11. This small team formed and dispersed quickly. They needed to work fast, efficiently, and effectively. Practically applying mission command, the team developed a framework to engage Queensland's leadership, an engagement vital in creating a plan for transitioning Queensland from crisis to reconstruction. Ultimately, through the judicious employment of mission command, that group of ADF planners team helped in guiding an entire Australian state's successful response to one of the country's most devastating disasters.

ENDNOTES

1   The team of seven planners was led by Lieutenant Colonel
    (LTCOL) Jim Hammett (Royal Australian Infantry Corps),
    and included LTCOL Sue Graham (Royal Australian Corps
    of Transport), Captain Evan Armstrong (Royal Australian
    Corps of Signals), Lieutenant Commander Jo Beadle,
    Royal Australian Navy, and Squadron Leader Alan Brown,
    Royal Australian Air Force. In addition, Queensland Police
    Service members, who are also Army Reserve Officers,
    made considerable contributions to planning, to include:
    Superintendent Mark Plath (Colonel, Army Reserve) and
    Detective Senior Sergeant Steve Vokes (Lieutenant Colonel,
    Army Reserve). In addition, the Chair of the Queensland
    Reconstruction Authority was Major General Mick Slater and
    the Chief of Operations and Plans was Colonel Chris Field
    (the author).

2   Australian Army Journal, *Point Blank—An Interview with
    Brigadier Mick Slater, Commander JTF 631*, Volume III, Number
    2, Winter, 2006, 11.

3   Commonwealth of Australia, Australian Army, *Land Warfare
    Doctrine 1, The Fundamentals of Land Power*, Canberra 2014,
    45; and Headquarters Department of the United States Army,
    Army Doctrine Reference Publication 6-0, Mission Command,
    Change No. 2, Washington, DC, 28 March 2014, 2.1.

4   Frederick P. Brooks Jr., *The Mythical Man-Month: Essays on
    Software Engineering*, Anniversary Edition (2nd Edition), 12
    August 1995. Brooks' Law refers to a software development
    principle coined by Frederick Brooks in *The Mythical Man-
    Month*. The law, "adding manpower to a late software project
    makes it later," concludes that when a person is added to a

project team, and the project is already late, the project time is longer, rather than shorter. The cause of these delays are: Increased "ramp up" time is required for new project members because of the complex nature of software projects. Such ramp up time takes existing resources (productive personnel) away from active software development, putting them in training roles.

Increases in staff size complicate communications between members, e.g., by increasing the number and variety of communication channels.

5   Central Intelligence Agency, *The World Factbook*, https://www. cia.gov/library/publications/the-world-factbook/geos/as.html (accessed August 26, 2015).

6   Queensland Government, *Interesting facts about Queensland*, https://www.qld.gov.au/about/about-queensland/statistics-facts/facts/ (accessed August 26, 2015).

7   The State of Queensland, Queensland Treasury, Queensland Government, *The Queensland economy, Strengthening our sectors*, https://www.treasury.qld.gov.au/economy/the-queensland-economy/index.php (accessed 06 September 2015)

8   Queensland Government, *Department of Community Safety Annual Report 2010-11*, Brisbane, February 2013, 3, http://www. disaster.qld.gov.au/Disaster-Resources/Documents/SDMG%20 Annual%20Report%202010-11%20Final.pdf (accessed 05 October 2015).

9   Ibid.

10   The Australian Honours Secretariat, *National Emergency Medal*, https://www.gg.gov.au/australian-honours-and-awards/ national-emergency-medal#qld (accessed August 26, 2015).

11 Queensland Government response to the Floods Commission of Inquiry Interim Report, August 2011, 1, http://www. emergency.qld.gov.au/documents/qld%20gov%20response-to-flood-inquiry.pdf (accessed August 26, 2015).

12 Queensland Government, *Operation Queenslander, The State Community, Economic and Environmental Recovery and Reconstruction Plan 2011–2013*, 23 March 2011, 2.

13 Ibid. For more information on Commonwealth-State Natural Disaster Relief and Recovery Arrangements, see: Australian Government, Attorney-General's Department, Emergency Management in Australia, Natural Disaster Relief and Recovery Arrangements. http://www.disasterassist.gov.au/ NDRRADetermination/Pages/default.aspx (accessed August 26, 2015).

14 In 2011, the Australian government's policy in this regard was outlined in *Defending Australia in the Asia Pacific Century: Force 2030 (Defence White Paper 2009).*

15 Commonwealth of Australia, *Defence White Paper 2009 — Defending Australia in the Asia Pacific Century: Force 2030*, Canberra, 2009, 41.

16 Commonwealth of Australia, *Defence White Paper 2009 — Defending Australia in the Asia Pacific Century: Force 2030*, Canberra, 2009, 11, 62.

17 Commonwealth of Australia, *Defence White Paper 2009 — Defending Australia in the Asia Pacific Century: Force 2030*, Canberra, 2009, 24.

18 Australian Army, *Operation QUEENSLAND FLOOD ASSIST*

2011, 15 March 2012 http://www.army.gov.au/Our-work/ Community-engagement/Disaster-relief-at-home/Operation- QUEENSLAND-FLOOD-ASSIST-2011 (accessed August 26, 2015); and Department of Defence, *Operation Queensland Flood Assist*, Media Release, 05 February 2011 http://www.defence. gov.au/media/departmentaltpl.cfm?CurrentId=11386 (accessed August 26, 2015).

19 Australian Army, *Operation YASI ASSIST 2011*, 29 May 2014, http://www.army.gov.au/Our-work/Community- engagement/Disaster-relief-at-home/Operation-YASI- ASSIST-2011 (accessed August 26, 2015); and Department of Defence, *Defence Responds to Tropical Cyclone Yasi*, Media Release, 02 February 2011 http://www.defence.gov.au/media/ departmentaltpl.cfm?CurrentId=11366 (accessed August 26, 2015).

20 Commonwealth of Australia, *Defence White Paper 2009— Defending Australia in the Asia Pacific Century: Force 2030*, Canberra, 2009, 11, 62.

21 Queensland Government, *Operation Queenslander, The State Community, Economic and Environmental Recovery and Reconstruction Plan 2011–2013*, 23 March 2011, 14.

22 Queensland Government, *Operation Queenslander, The State Community, Economic and Environmental Recovery and Reconstruction Plan 2011–2013*, 23 March 2011, 15 http:// qldreconstruction.org.au/u/lib/cms2/operation-queenslander- state-plan-full.pdf (accessed August 26, 2015).

23 A campaign is a series of simultaneous or sequential operations designed to achieve one or more strategic objectives. Commonwealth of Australia, Australian Army, *Land*

Warfare Doctrine 1, The Fundamentals of Land Power, Canberra 2014, 19.

24 Jonathan Bailey, Richard Iron & Hew Strachan (eds.), *British Generals in Blair's Wars*, Chapter 9, *Modern Campaigning: From a Practitioner's Perspective, General (Retired) Sir John McColl*, Ashgate Publishing, Surrey, 2013, 114.

25 Queensland Government, *Operation Queenslander, The Community, Economic and Environmental Recovery and Reconstruction Implementation Plan 2011-2013*, 29 April 2011 http://qldreconstruction.org.au/u/lib/cms2/operation-queenslander-implementation-plan-full.pdf (accessed August 26, 2015); Queensland Government, *Operation Queenslander, A Guide to Local Community, Economic and Environmental Recovery and Reconstruction Planning*, 29 April 2011 http://qldreconstruction.org.au/publications-guides/reconstruction-plans/local-plan (accessed August 26, 2015).

26 Headquarters Department of the United States Army, *Army Doctrine Reference Publication 6-0, Mission Command*, Change No. 2, Washington, DC, 28 March 2014, 2.1.

27 Headquarters Department of the United States Army, *Army Doctrine Reference Publication 6-0, Mission Command*, Change No. 2, Washington, DC, 28 March 2014, 2.1.

28 Commonwealth of Australia, *Defence White Paper 2009— Defending Australia in the Asia Pacific Century: Force 2030*, Canberra, 2009, 11, 62.

29 Commonwealth of Australia, *Defence White Paper 2009— Defending Australia in the Asia Pacific Century: Force 2030*, Canberra, 2009, 24.

30 Commonwealth of Australia, *Defence White Paper 2009 — Defending Australia in the Asia Pacific Century: Force 2030*, Canberra, 2009, 59.

31 The six Queensland Government departments were: Queensland Department of Communities (DoC), Department Employment Economic Development and Innovation (DEEDI), Department of Environment and Resource Management (DERM), Department of Transport and Main Roads (DTMR), Department of Public Works (DPW), and the Department of the Premier and Cabinet (DPC).

32 Operation Queenslander six lines of reconstruction: Human and Social; Economic; Environment; Building Recovery; Roads and Transport; and Community Liaison and Communication.

33 Headquarters Department of the United States Army, *Army Doctrine Reference Publication 6-0, Mission Command*, Change No. 2, Washington, DC, 28 March 2014, 2-2.

34 See Australian Defence Force Publication (ADFP) 5.0.1 — *Joint Military Appreciation Process*, Edition 2, 24 February 2015.

35 Queensland Government, *Operation Queenslander, The State Community, Economic and Environmental Recovery and Reconstruction Plan 2011–2013*, 23 March 2011, 10.

36 Headquarters Department of the United States Army, *Army Doctrine Reference Publication 6-0, Mission Command*, Change No. 2, Washington, DC, 28 March 2014, 2-3.

37 Cable News Network, *Mandela at 90 — Mandela in his own words*, June 26, 2008, http://edition.cnn.com/2008/WORLD/africa/06/24/mandela.quotes/ (accessed 11 October 2015).

38 Nelson Mandela, *Long Walk to Freedom: The Autobiography of Nelson Mandela Paperback*, Back Bay Books, October 1, 1995, 22.

39 Dwight D. Eisenhower, 235—Speech at the National Defense Executive Reserve Conference, Washington, D.C., November 14, 1957, http://www.presidency.ucsb.edu/ws/?pid=10951 (accessed 06 September 15).

40 Queensland Government, *Operation Queenslander, The State Community, Economic and Environmental Recovery and Reconstruction Plan 2011–2013*, 23 March 2011, 14. http://qldreconstruction.org.au/u/lib/cms2/operation-queenslander-state-plan-full.pdf (accessed August 26, 2015).

41 *Toward Q2: Tomorrow's Queensland* was released in September 2008 as the government's vision for Queensland. It outlines five ambitions and ten long-term measurable targets for a strong, green, smart, healthy, and fair Queensland by 2020. The State of Queensland, Department of the Premier and Cabinet, Queensland Government, *Toward Q2 2011-2012 Target Delivery Plans* http://www.cabinet.qld.gov.au/browse.aspx?category=Q2 (accessed 06 September 2015).

42 Headquarters Department of the United States Army, *Army Doctrine Reference Publication 6-0, Mission Command*, Change No. 2, Washington, DC, 28 March 2014, 2-4.

43 Queensland Government, *Interesting facts about Queensland*, https://www.qld.gov.au/about/about-queensland/statistics-facts/facts/ (accessed August 26, 2015).

44 Queensland Reconstruction Authority, http://qldreconstruction.org.au/publications-guides/reports/monthly-reports [accessed 27 September 2015]. For more information on the legislation that formed the Queensland

Reconstruction Authority, see the Queensland Government, *Queensland Reconstruction Authority Act 2011*, Act No. 1 of 2011, http://www.legislation.qld.gov.au/LEGISLTN/ACTS/2011/11AC001.pdf (accessed 27 September 15).

45 United States Army, Field Manual 100-5, *Tentative Field Service Regulations — Operations*, Oct 1939, 62.

46 Headquarters Department of the United States Army, *Army Doctrine Reference Publication 6-0, Mission Command*, Change No. 2, Washington, DC, 28 March 2014, 2-4.

47 Headquarters Department of the United States Army, *Army Doctrine Reference Publication 6-0, Mission Command*, Change No. 2, Washington, DC, 28 March 2014, 2-4.

48 Michael Howard & Peter Paret, *Carl Von Clausewitz, On War*, Princeton University Press, 1989, 102. (Emphasis in original)

49 Headquarters Department of the United States Army, *Army Doctrine Reference Publication 6-0, Mission Command*, Change No. 2, Washington, DC, 28 March 2014, 2-5.

50 Queensland Government, *Operation Queenslander, The State Community, Economic and Environmental Recovery and Reconstruction Plan 2011–2013*, 23 March 2011, 10.

51 Queensland Government, *NDRRA information for applicants*, Queensland Reconstruction Authority, Brisbane, Queensland, 2015 http://qldreconstruction.org.au/ndrra (accessed 05 October 2015).

52 Queensland Government, *Submission Productivity Commission Inquiry Natural Disaster Funding Arrangements*, Brisbane, June

2014, 1, http://www.pc.gov.au/inquiries/completed/disaster-funding/submissions/submissions-test/submission-counter/sub031-disaster-funding.pdf (accessed 05 October 2015). The Australian and US dollars were virtually on par in January 2011. See "Australian Dollars (AUD) to US Dollars (USD) exchange rate for January 31, 2011," http://www.exchange-rates.org/Rate/AUD/USD/1-31-2011, (accessed October 30, 2015).

53 Commonwealth of Australia, Australian National Audit Office, *Audit Report No. 8 2013–14 The Australian Government Reconstruction Inspectorate's Conduct of Value for Money Reviews of Flood Reconstruction Projects in Queensland*, Barton, Australian Capital Territory, 06 November 2013, 41, http://www.anao.gov.au/~/media/Files/Audit%20Reports/2013%202014/Audit%20Report%208/AuditReport-2013-2014_08.pdf (accessed 05 October 2015).

54 Commonwealth of Australia, Australian National Audit Office, *Audit Report No. 8 2013–14 The Australian Government Reconstruction Inspectorate's Conduct of Value for Money Reviews of Flood Reconstruction Projects in Queensland*, Barton, Australian Capital Territory, 06 November 2013, 41, http://www.anao.gov.au/~/media/Files/Audit%20Reports/2013%202014/Audit%20Report%208/AuditReport-2013-2014_08.pdf (accessed 05 October 2015).

55 Commonwealth of Australia, Australian National Audit Office, *Audit Report No. 8 2013–14 The Australian Government Reconstruction Inspectorate's Conduct of Value for Money Reviews of Flood Reconstruction Projects in Queensland*, Barton, Australian Capital Territory, 06 November 2013, 41, http://www.anao.gov.au/~/media/Files/Audit%20Reports/2013%202014/Audit%20Report%208/AuditReport-2013-2014_08.pdf, (accessed 05 October 2015).

56 Headquarters Department of the United States Army, *Army Doctrine Reference Publication 6-0, Mission Command*, Change No. 2, Washington, DC, 28 March 2014, 2-5.

13

THE 21ST-CENTURY STATE OF PLAY: THE APPLICATION OF MISSION COMMAND IN AUSTRALIA'S 2015 ARMY COMBAT BRIGADE

Major General Roger Noble, DSC, AM, CSC, Australian Army

Exercise HAMEL 2015,
[A two-sided, free play, formation level exercise]
Shoal Water Bay Training Area, Queensland, Australia
XX2350hrs Jul 15

Commanding Officer, 3rd Battalion Royal Australian Regiment Battle Group (CO 3 RAR): *A Company has now been cut off forward. They have a battle group minus to their rear. There is not much I can do to support them from the main defensive position. They either stay as we planned or breakout to rejoin to the main position now. The latter will likely take longer than we war-gamed — longer route, tougher ground.*

Commander, 3rd Brigade [COMD 3 Bde]: *No. It is different than the ROC [rehearsal of concept] drill. We know the enemy will pause now — it is a pattern. Move, stop, move, stop. There is a window to attack them. Tell him [OC A Coy] to attack now, back towards you. I have nothing more to give you; go with what you've got.*

CO 3 RAR: *He has at least seven enemy tanks behind him, a battle group to his west, and unknown forces to his front. I am not sure that is going to be viable.... (One of those difficult pauses) ... What do you really want him to achieve?*

COMD 3 Bde: *Maximum disruption as the enemy pauses. Attack by night when they least expect it. Force them into another planning cycle. Drive them off balance, induce delay, let them know we will go for them when they least expect it.*

CO 3 RAR: *How about I tell him that and let him decide? You know he will get it done.*

. . . . and he did. Following the "old school" landline conversation above, the officer commanding A Company 3 RAR attacked in the opposite direction to that I expected to the complete surprise of the enemy who were, it turns out, completely unaware of his presence. He surprised both of us. As ever, friction and uncertainty is an issue for the foe just as much as it is for the good guys. This simple exercise example reinforced what we have known since Gallipoli: the capacity for relatively young officers to understand intent and seize fleeting opportunities is critical to tactical success. The primary lesson for me was that the higher you go the more you seem to rely on this capacity and their courage and skill. You can't do it yourself. It is as close to a military scientific law as you can get that an Army founded on a philosophy of mission command and armed with a bias for action must remain the enduring institutional aspiration, especially for the forces of smaller, democratic nations that deeply value the lives of their soldiers and for which mass is no viable alternative.

This book has reviewed mission command in the Australian context over a wide variety of circumstances and at different times. This concluding chapter is aimed squarely at the present and future. Based primarily on my experience as a combat brigade commander, immersed in the multiple changes and challenges facing the contemporary force, it seeks to capture the mission command state of play inside the Australian Army through a series of observations. These are drawn from my reflections on a variety of recent experiences—operational, during training and in terms of force modernization. It also looks forward and suggests ideas and issues to suggest how we should focus in our quest to develop an effective 21st-century mission command climate supported by an organization that fosters and grows the philosophy at every opportunity.

ESSENTIAL CONTEXT

Like all Western armies, especially those having time in Iraq and Afghanistan, the Australian Army has evolved rapidly over the past fifteen years. It is critical to have a basic understanding of this journey of from the perspective of the combat brigade during this hectic period.

Those last fifteen years have seen a quiet almost-revolution in the Australian Army, during which the combat brigades have played a central role. The primary drivers behind this change include the pressure generated by the requirement to mount and rotate forces for multiple operations, need to rapidly adapt and modernize to meet the challenge of those operations, and the enduring requirement to optimally employ finite resources. Establishment of the combat brigade as the army's training and force generation hub has been central to the change. Scarce, critical, high cost, or sensitive resources such as helicopters and intelligence assets have been reorganized into three enabling brigades whose assets regularly reinforce the army's three combat brigades. The resulting reinforced combat brigade is able to employ a diverse range of capabilities not previously available at brigade level and below. This profound organizational change has been both driven by and takes place within a context of continuous modernization stemming from the lessons of recent conflicts; identification of new ways of doing business; and introduction of new technologies, systems, and capabilities. This renaissance of the combat brigade remains a dynamic, rolling work in progress.

It is critical to note that this is not a uniquely Australian experience. All western militaries have undergone similar evolutions in their own ways and contexts. The consequent turbulence is not new. The rate of change, however, especially that technological, is at least as great as at any previous time.

OBSERVATION 1: MISSION COMMAND MUST BE A FOUNDATION PHILOSOPHY THAT DRIVES ALL ACTION AND THINKING

Commitment to mission command cannot be mere rhetoric or a method applied only in tactical settings. It must be applied in all circumstances from battle group attack to running the base charity open day. Our collective understanding of what mission command means and commitment to it in all circumstances must be genuine. This is an issue that extends well beyond tactical scenarios and the battlefield. It should not surprise if there is an absence of tactical flare and failure to seize initiative if soldiers and officers are nurtured in a compliance-driven, risk-averse, checklist environment of endless detailed regulation, audit, and supervision.

At its heart, mission command requires the mission— *the purpose*—to have primacy. It demands that *all* find ways of achieving the designated outcome within defined constraints and requirements when given adequate resources. The mission is the binding agent and center around which all spins. Requirements and constraints are not the purpose and end in themselves; the purpose has primacy and must drive all action. Does the modern Australian Army get this? How is it fairing in the contemporary combat brigade?

Employment of mission command is unequivocally a foundation belief within the contemporary combat brigade. Nobody talks any alternative (even when they do not adhere to the basic tenets of mission command!). In theory, mission command is dominant. In practice, while intent-based direction and the use of mission orders is the norm, there are leaders who exercise an extent of direct control not in keeping with the philosophy.[1] There is a common mission command understanding directly reflected in extant combat brigade battle procedures[2] and across all corps[3] down to the lowest levels. Critically, for the first time in living memory, brigade and other maneuver unit standard operating procedures (SOPs) are common across the entire force and subject to systematic annual review. This includes systematic audit through a series of collective training activities delivered by the Combat Training Center culminating in the annual Exercise HAMEL, which evaluates the reinforced combat brigade during a two-sides free play test. As a result, mission command is embedded in the lexicon, psyche, doctrine, procedures, and evaluation of contemporary combat brigades. Quite simply, we believe in it.

Our principal weakness relates to the strength of our common understanding regarding what we mean by "mission command." This is our greatest source of friction. The biggest variance concerns different interpretations of the level of freedom of action that should be devolved. Some people think mission command means specific detailed direction should never be mandated by a higher commander, that freedom of action is always given to subordinates: "Don't tell me how to do it, just tell me what you want done." The absence of an alternative model to mission command and relatively unsophisticated discussion of control in Australian doctrine combine to provide limited guidance in a way avoided in more comprehensive US command and control (C2) doctrine.[4] It is clear that specific direction and supervision are not inconsistent with a mission command philosophy. However, we need to better teach our people about the limits of mission command and alternative ways of doing business.

Our doctrine needs some work. On the plus side, it already mandates mission command as the organizational command and control philosophy. Australian Army doctrine at every level enshrines intent-driven orders and a maneuver-driven planning process. Mission command as defined is consistent with professional western "best practice." The weakness is the absence of a comprehensive command and control philosophy with logic similar to USMC or recent US Army doctrine. Our current command, leadership, and management doctrine reads more like academic leadership theory than operational command and control doctrine. No Australian doctrine articulates an alternative to mission command such as *detailed control*.[5] We all know the detailed control guys when we see them in action, but we do not have a formal doctrinal construct to explain this approach and when it is justified. In the absence of an articulated alternative, our ability to understand what constitute the limits of effective mission command is challenged. Alignment between army and joint doctrine is also imperfect. An audit to align all doctrine alongside a rewrite of our foundation C2 doctrine would be timely.

OBSERVATION 2: A RELENTLESS BIAS FOR ACTION FOUNDED ON SYSTEMATIC RISK ASSESSMENT (NOT RISK AVERSION) MUST BE ROUTINE AND ACCEPTED ACROSS THE FORCE.

If subordinates won't or are not permitted to take calculated risks and act boldly to seize the initiative, then a genuine mission command culture does not exist. Evidence suggests that our junior commanders have a bias towards action but struggle to know exactly when and how to exercise it. This is normal; it takes experience and practice to refine their decision-making in the face of uncertainty. There were a number of striking examples in this regard during Exercise HAMEL 15: commanders at combat team level and below deviating from their directed tasks to achieve the designated purpose. They were not always completely successful, yet the outcomes were accepted and—most importantly—the behavior was strongly reinforced.

A bias for action only exists where a considered tolerance for "mistakes" or poor outcomes also exists. This tolerance for action and mistakes made in accordance with intent is actually quite high—at least at the tactical combat brigade level and below. A combination of doctrine, operational experience, education, and the recent emphasis on two-sided free play exercises has profoundly reinforced our understanding of the fog of war, that the enemy is an independent actor, and that friction pervades all. Chance is an enduring reality with which we must all deal. We know—because we have both studied it and seen it—that the capacity to act when operating within an uncertain and complex environment is critical to shaping the environment, the enemy, and achieving mission success.

During the exercise of mission command in a tactical setting, mistakes are readily accepted and then dissected as learning opportunities via the After Action Review (AAR) process imported from the US military during the 1990s.[6] Americans are nothing if not systematic and our selective use of their tools and ideas has been a great benefit. Planning and back-briefing processes at battle group, brigade, and higher echelons allows for actions to be taken in accordance with intent given mandated restrictions and constraints. An emerging organizational approach for both tactical and non-tactical environments is evident in the following extract from the current 3rd Brigade organizational intent statement:

A bias for action: Individuals and organizations must act decisively and confidently IAW [in accordance with] the commander's intent. When a problem emerges it should be fixed at the lowest level possible. When it can't be fixed by the individual or unit present, find someone who can fix it and/or elevate quickly. Action in the real world means mistakes occur. We can't control everything and we shouldn't try to. Mistakes are expected. They are essential learning opportunities and not to be feared. A breach of intent is, however, an unacceptable error as it undermines trust and therefore destroys the confidence that underpins mission command. I will not tolerate breaches of intent or operate with a lack of trust. Nor will I accept inaction unless it is a conscious and calculated decision not to act taken IAW the commander's intent.[7]

It remains arguably our most difficult training challenge to inculcate a willingness toward action and develop the judgment to know when and how to act when facing uncertainty. The steadily increasing frequency of individual and collective force-on-force training combined with input from purpose-designed organizations such as the Combat Training Centre (CTC) are increasingly putting commanders in demanding "post-H-hour" scenarios where intent becomes the rope in the dark and accepting the need to act during uncertainty becomes essential.

Learning by doing is essential and our opportunities to do so have been significantly expanded. When asked in 2015 to identify their top tactical challenges, the majority of combat team commanders at brigade and below echelons identified the practical aspects of decision-making and execution in accordance with their commander's intent in a fluid, uncertain environment. The challenges included (1) analyzing the mission, (2) surveillance and reconnaissance planning and execution, and (3) tactical language and task verb selection.[8]

There is no better way to educate a force on the need for mission command than to put its senior commanders into a complex scenario characterized by chance, friction, and uncertainty. For the Australian Army, this is now "army normal."

OBSERVATION 3: IT IS ALL ABOUT THE PEOPLE AND HOW THEY WORK TOGETHER TO EXERCISE COMMAND AND ACHIEVE EFFECTIVE CONTROL

We have found we always return to the same critical issue: humans and how they think, work, and interact. This is the key no matter the dazzling technology, time period, or the operational circumstances.

The commander is certainly a key player, but there are many commanders in any combat brigade. *How well they understand and trust each other and how effective they are in turning intent into coherent, timely action is what ultimately matters.* This is certainly not a new idea, but it is a profoundly important one easily lost in the frantic reality of the information age. Armies that do not embrace this salutary point will struggle to be effective in mission command exploitation.

As a force, we unequivocally understand command is a human activity. You need only two things to command: humans and vested authority. You sometimes don't need much to exercise command; a look can do it. Face-to-face and verbal communications are therefore an essential and enduring requirement underpinning effective command. Emails, Battle Management System messages, online chat, and other remote means of communication just don't cut it and never will, especially if people are shooting at you and those you command. How you express intent and how well it is understood are critical. It is the commander's responsibility to ensure that he or she is understood, the intent is communicated and constantly updated, and a control framework exists so that intent can be translated into action. The modern effective commander must be a communicator who owns the whole problem set and implicitly accepts complete responsibility for organizational performance. The traditional view of a (Commonwealth) general reclined in period furniture; draped in immaculate uniform of Sam Browne belt, red tabs; and operating via stony and remote direction won't fly in the 21st century—if it ever did.

Whether at section, battle group, or task force level, commanders must make decisions. They have to express intent clearly and give directions that must then be executed and interpreted by subordinates who operate in chaotic, uncertain, violent, and ever-changing operational environments. Subordinates need to believe in what they are doing and those telling them to do it, especially when they face death or injury. Leadership is therefore fundamental to effective command. This is in keeping with our doctrine and fundamentally reinforced by our recent experiences.

The real 21st-century challenge is in understanding control and how we establish a system to enable effective action in accordance with commanders' intents. There are multiple nuances that directly impact the effective execution of mission command.[9] Decisions and the plans that frame them need to be executed in a complex, multi-level environment which is inherently chaotic. Multiple disparate actions by many dispersed actors ideally need to be reinforcing even if not closely synchronized. While commanders can and do exercise control, it is action executed through multiple commanders and staffs bound by a common mission, intent, and control measures that ultimately matters. In a combat brigade, the more standardized, flexible, and simple the control systems available (technology, process, education and training), the more likely they will be successful. We have found this is a difficult challenge to master in practice given dynamic operating environments and diversity of capabilities found in the modern combat brigade. It is easy to talk control but much harder to exercise it effectively.

Australian reinforced combat brigades have learned or re-learned that how we establish a control system—*who* does *what*, *how* it is done, and *when* it is done—is the critical C2 challenge. We have learned yet again that how a staff operates is critical, especially in the modern world of multiple specialists and capabilities. The primary lesson learned can be summarized in the simple equation:

Best Chance of Success =
Mission focus + Discipline + Teamwork + Quality staff.

At the heart of any effective headquarters is a set of robust and proven standard operating procedures (SOP) executed by talented, knowledgeable individuals who put the mission *above all else*—especially themselves. Given the structure and size of our army, it is now unavoidable that we have a single set of combat brigade SOP. We must also retain the capacity to vary from the standard guidance when mission and circumstances require.

Discipline, and more importantly self-discipline, is also vital. We have all observed large headquarters working 24/7, generating endless product. We have learned time and again that less is more, that quality, not quantity, matters in providing guidance. Staff actions and processes must serve commanders, those senior, immediate, and subordinate. As Commander 3rd Brigade, I expressed this as the critical characteristic of self-discipline which was articulated as follows:

Professional soldiers beat warriors and amateurs every time. The conduct, actions, capacity, and behaviour of the individuals within the brigade will ultimately determine its quality. Self-discipline, self-awareness, and personal humility are the signature traits of the professional soldier and these characteristics distinguish him or her from "the warrior." The professional soldier of any rank is a thinking decision-maker who is highly skilled, unafraid to act, and always does so IAW the commander's intent. He or she automatically operates inside the framework of our core beliefs—no other way is possible. It is a command responsibility to set the conditions for all of us to operate at this level—self-discipline must be taught, explained, demonstrated, lived, and demanded. Individual self-discipline underwrites individual, organizational, and, ultimately, institutional resilience and adaptability.[10]

Teamwork—and a ruthless collective and individual commitment to it—is also vital. Again, this is not a new idea. There is (or should be) little individual glory to be had within the headquarters environment where the staff officer is always, in the words of General Sir John Monash, "the servant of the troops." The modern brigade headquarters contains soldiers and officers of every hue, skill, and background. It is also a friction-filled environment; the only way to overcome this is a disciplined commitment to work together regardless of personality or personal interests. Negative internal competition and personal frictions cannot be tolerated; everyone needs to understand this. *Courage, endurance, mateship* and *sacrifice*—words etched on the black rocks above Kokoda—are just as relevant to the 21st-century staff team as they were to the Diggers fighting for their lives in 1942.

OBSERVATION 4: TECHNOLOGY IS AN OPPORTUNITY NOT A "SILVER BULLET"; HOW WE USE IT IS WHAT REALLY MATTERS.

Since we recognize that equipment is but a means to an end and not the end itself, our doctrine is independent of any particular technology.[11]

General (USMC) Charles C. Krulak

General Krulak nailed it twenty years ago, but we seem to keep relearning this lesson time and time again. Technology gets plenty of attention and is often the headline when command and control are discussed. Digitization is a hot topic of military debate. The danger of being bedazzled remains high. Marines unsurprisingly follow up General Krulak's observation as follows:

There are two dangers in regard to command and control equipment, the first being an overreliance on technology and the second being a failure to make proper use of technological capabilities. The aim is to strike a balance that gets the most out of our equipment and at the same time integrates technology properly with the other components of the system.[12]

Does the Australian combat brigade have an effective and capable technical system that supports mission command? How is today's brigade handling the flood of technological change? Has it struck a proper balance?

A number of definitive statements can be made based on our recent experiences with technology. The evolution of command, control, communications, computers, and intelligence (C4I) technology is unlikely to stop; constant upgrade, replacement, and redesign are likely constants.[13] We have never had better and more capable technology at the tactical level than at present.

There is no doubt that mission command is enabled by current technology. However, the current systems can also facilitate detailed control and micromanagement to a level never before possible. General Krulak's observation at the opening of this section still holds true. Now—as ever before—it is how the technology is employed in conjunction with a belief system that determines the command approach.[14]

Careful and close consideration of how we use our technology is, therefore, a must. It isn't just the S6's (staff communications section's) bag. There is no question that modern technology allows the application of the 29,000km screw driver from echelons above the brigade or from within it. This power is useful and is not inconsistent with mission command when employed to check, inform, gain feedback or direct at critical points. It can, however, be a genuine threat to freedom of action when employed as a blind default and without thought—because we can and it makes someone feel in control. A far more commonly observed error is to not fully understand that how we use our tools directly and profoundly impacts on how command is exercised in practice. Commander's unconnected with or ignorant of our contemporary means of communication and how they are used have often found themselves doing what *can* be done rather than what *should* be done. Technology and the tools, therefore, demand understanding by the commanders who must also own how they are used and how this impacts on subordinates and superiors alike. Given the rate of change and the tremendous scope of available technological options, this is now a fundamental component of contemporary professional mastery for commanders to understand and employ in shaping how the tools are used. General Krulak definitely nailed it.

OBSERVATION 4: A HARD REALITY—IT IS IN WITH "THE NEW" BUT NOT OUT WITH "THE OLD"

This observation follows directly from our experience using new systems tactically against enemies with capabilities similar to our own. For all the opportunities offered by new systems, they do not necessarily replace the old approaches but rather augment them—which means *more* work, not less, for those employing them. We still need to be able to fight in the dark, without power, and without GPS in hostile, limited communications environments against foes that directly target our C2 and other technological vulnerabilities. Redundancy remains essential. It is therefore important for battle staff and commanders to retain and maintain "analog" and traditional skills both to exploit enemy vulnerabilities and deal with enemy action against our new systems.

OBSERVATION 5: THE SIMPLE PLANS VIOLENTLY EXECUTED RULE

Clausewitz made an important observation long before the existence of battle management systems, data transfer, electronic warfare and most other tools of modern war: "Everything in war is very simple, but the simplest thing is difficult."[15] The contested 2013-15 HAMEL exercise series pitting us against an enemy with capabilities near our own taught me that this has not changed one bit.

Simple does not mean unsophisticated. Simple means clear and coherent in a way that can be readily understood by all. The power of a plan that makes sense to generals, soldiers, and everyone in between is frankly difficult to match. Again, humans are the key. A plan based on clear, critical thinking linked to a set of assumptions subject to systematic verification of their continued validity is a unifying core for focusing mission command-led action in the most complex of environments. Complex systems theory has given rise to the important notion of *self-organization* by sub-elements within a complex system. A clear, simple plan executed by a disciplined, capable force that operates within a mission command philosophy allows all players to make decisions that aggregate towards mission success. Such a force becomes a military mission command-enabled self-organizing system.

Violent execution is not limited to the application of violent means. Forthright execution of a plan using all available means has a profound impact on all involved, especially the enemy. Action shakes the tree and reorders the system. The great tendency is often to wait for that next key piece of information or the final piece of the puzzle. Generals to corporals feel this pressure, especially in an age where there are so many tools available to answer our questions. On recent operations, we have faced adversaries who were ruthless, clever, and adaptive but often limited in the means that they can employ [e.g., they rarely had modern air or intelligence, surveillance, and reconnaissance (ISR capabilities)]. We often (but not always) have had time to wait and think and it is has been hard for them to punish us inside our well defended forward operating bases given our air supremacy. Our recent two-sided, exercises against forces armed with systems like our own revealed just how pressed we are for time when confronted with a near-peer competitor. We repeatedly experienced how hard it is to determine enemy intent and dispositions despite our capabilities, especially in complex terrain. These experiences reaffirm the power of taking action in accordance with the commander's intent at every level. From this comes the need to improve this willingness across the entire force.

In the past ten years the Australian Army has reaffirmed the need to be able to develop simple, clear plans violently executed by trained professional soldiers. The sad but important conclusion is that gadgets alone cannot overcome the play of chance and do not eliminate friction or uncertainty…at times they only add to the friction.

OBSERVATION 6: INSTITUTIONAL CULTURE MATTERS AND NEEDS NURTURING

Culture matters, even if it is hard to find an agreed definition of just what culture actually is. Without delving into anthropology, let us agree that culture involves a complex mix of factors resulting in a belief system based on a combination of stated and unstated assumptions. Australian Army capstone doctrine requires that we develop and foster a mission command climate. Ours are, however, not mission command organizations just because we say we are. A critical question must therefore be whether our contemporary institutional culture is supportive of mission command. Can it tolerate "mistakes?" Will it allow independent action by subordinates? Will it permit subordinates to disobey orders to achieve the mission?

So how do we stand on the culture front? Australian Army culture, on balance, appears to be well suited to fostering an effective and robust mission command culture. For a start, the army assigns particular importance to its institutional culture; it is a key focus in the latest capstone doctrine publication.[16] The core values of the army—*courage, teamwork, initiative,* and *respect*—appear entirely consistent with building the foundation of trust so essential to mission command. Analysis of 3 Brigade over the period 2012-2015 indicates an unequivocally increasing adherence to the core values of the organization.[17] Mission command is genuinely the C2 method to which the organization aspires. Impediments are regularly discussed in both the tactical and barracks environments. Our traditions and history focus on the individual digger (Australian soldier). Independent initiative and action-willingness are highly prized. There are other important elements of army culture such as a faith in detailed planning, professional competence, and combined arms excellence that are also consistent with a mission command culture. It is not an accident that today's key collective annual training event is named after the 1918 battle of Hamel, John Monash's carefully planned and executed combined arms set-piece success. The century-plus mix of history, stories, and myth regarding the Australian Army are a strong "fit" with the requirements of effective 21st-century mission command.

Joint culture is perhaps even more difficult to assess and define. Some might challenge the existence of a genuine Australian joint operational culture. Recent operations in the Middle East and Central Asia, while executed by joint task forces, are more accurately described as a series of relatively discrete single service tactical contributions to combined joint coalition operations than Australian joint undertakings. In contrast, the country's ongoing amphibious capability development requires forging of a practical joint approach. It has revealed the requirement to identify and apply innovative ways of genuinely conducting joint business. There is also no denying the consistent, positive, pragmatic, and collaborative approach adopted by Australian Defence Force (ADF) elements when grouped together on operations: they work it out together. This was confirmed during recent domestic and offshore humanitarian relief operations.[18] However, that there exists a durable, systematic joint culture of learning and review is questionable. For example, a 2011 external audit of the learning process by the Australian National Audit Office concluded the ADF approach to learning operational lessons was "patchy and fragmented."[19] While mission command is the doctrinally mandated ADF-wide approach to C2 and the stated principles are identical to the army doctrinal prerequisites, it is arguable whether it is the actual dominant method in any of the services.[20] Anecdotal evidence indicates a strong preference for detailed control or ad hoc methods some of which have been covered by other authors in this book. It is clear that a systemic analysis of service and joint C2 cultures and practice require review. Successful execution of mission command in combat brigades and other units is logically reliant on both service and joint C2 culture and practice.

Observation 8: We Have Learned That We Must "Learn by Doing"... And That This Never Ends

The Australian Army Combat Training Centre staff will tell you we are not perfect. We repeat mistakes. We need to work hard on applying mission command on the battlefield.[21] They are right, of course. It is understandable if the CTC observer trainer [OT] gets depressed during a posting as he sees the same mistakes repeated by different units. He or she should not be so disheartened. To quote the last words of Australia's greatest outlaw—Ned Kelly—from the scaffold, "Such is Life."

The central deduction to be made from the repetition of shortfalls is that mission command requires doing; it is not merely a theoretical, classroom activity. Mastery by one unit it does not automatically transfer to others even within a supportive and systematic learning system; it must be cultivated by commanders and teams. The truth is we will always disappoint observer trainers as new teams and individuals face the challenge of mastering our business. This is why the independent and systematic coaching and mentoring of a professional combat training center (and the wider system) is critically important as future generations learn by doing over and over again. The institution must capture and disseminate best practices. We must have an effective corporate memory.

We in the Australian Army have learned that if we throw high quality people into near-real test conditions and arm them with the latest capabilities they will innovate, learn, and inform us regarding the way ahead. An example of this in action can be drawn from the digitization of 3rd Brigade in 2014. Our dismounted battle management system had inherent limitations. The brigade's first tactical exercise involved an air mobile raid of over 80km distance. I entered the brigade tactical headquarters to monitor the mission, anticipating only our standard secure voice-enabled communications with the raid force. To my surprise the raiders appeared as "blue dots" on our battle management system network, allowing us to monitor their location in real time.[22] How could this be possible? The photograph below provides the answer. "A bunch of NCOs" took components of our digital system; safely and appropriately fitted them into an aging all-terrain vehicle; affixed a portable power source, and flew the resulting system forward. They took what we had and used it in a way no one had previously considered—smart, young soldiers of the digital age. It is almost an article of faith in our Army that the clever and capable can find solutions. The modern force needs this now more than ever and we should actively exploit this powerful strength. Inside the testing but supportive crucible of our contemporary force generation cycle, we have learned to take what we have and run with it. Only by doing can we truly learn; innovation is the priceless by-product.

Figure 13-1: Soldier Innovation—Adapting New Technologies to Operational Demands[23]

CONCLUSION

As a young officer, I proudly read in Norman Dixon's *The Psychology of Military Incompetence* that the Australian Army was given credit for inventing the term "bullshit"[24] in 1916 to describe and decry military "bull," the absence of logical thought and/or seemingly mindless actions of the professional armies encountered by our diggers in France. Given this important heritage, this paper has aimed to provide a "bullshit" free summary of the mission command state of play in the contemporary Australian Army combat brigade. This has been done through an articulation of a set of key observations drawn from my participation in our recent operational and training experiences.

It is clear the contemporary Australian combat brigade thinks and operates within a mission command philosophy. While its application is almost certainly imperfect, it is the genuine organizational aspiration and mandated requirement to command and control this way. It comprises almost a core belief in our army.

As noted in the opening of this chapter, the Australian Army has undergone an almost-revolution over the past fourteen years. The service has seen a revitalization of the combat brigade and significant investment in its capabilities, including command and control. Brigades—and the army in general—are arguably better enabled to execute mission command than ever before. Equally, a commander is better enabled to exercise detailed control. The vast majority of effort and thought is going into how to enable mission command. It is alive and well in practice. Given experience and opportunity, the modern Australian combat brigade should continue to refine and adapt its application over the years ahead. We now have a demanding force generation system built on contested free-play exercises involving near-peer adversaries. It is a system enabling us to continue testing, adapting, succeeding, or failing, ultimately to learn and innovate. We are not making it easy on ourselves and so it must stay.

This quest to nurture an effective mission command culture and professional system able to exploit it is never ending. Whether it is new technologies, new people, or new environments and missions, the messy, imperfect, and constant process of learning and innovating must continue. We will likely fluctuate in our effectiveness as a mission command organization. We must not rest, lose focus, or become complacent. That would be the quickest path to failure. No bullshit.

ENDNOTES

1   The structure of orders and mission statements and use of specified task verbs and agreed lexicon are mandated and commonly in use at all echelons both in the field and at training institutions.

2   *Reinforced Combat Brigade, Standard Operating Procedures* [SOP], Version 3.0, Army Knowledge Group, Tobruk Barracks, Puckapunyal, SOP 0-3, Foundation Concepts: Command and Control, n.d.

3   Examples of recent SOP development founded on manoeuvre theory and existing mission command concepts include *The Standard Infantry Battalion Handbook*, Vol.1, Platoon Operations, Part 2: Tactical Aide Memoir, 1st Addition, sections 1 and 2, April 24, 2015; and SOP—*The Standard Infantry Battalion Field Handbook*, Vol. 2: Battalion/Company Operations, 1st Addition, April 24, 2015, section 2—Command and Control.

4   The USMC proposes an alternate model employing a sliding scale of control delegation. Paraphrasing slightly, *detailed control* can be described as coercive, a term that effectively describes the manner with which the commander achieves unity of effort. He holds a tight rein, commanding by personal direction or detailed directive. Command and control tend to be centralized and formal. Orders and plans are detailed and explicit; their successful execution requires strict obedience and minimizes subordinate decision-making and initiative. Detailed command and control emphasizes vertical, linear information flow. In general, information flows up the chain of command; orders flow down. Discipline and coordination are imposed from above to ensure compliance with the

plan. United States Marine Corps, Marine Corps Doctrinal Publications (MCDP) 6, *Command and Control*, 77.

5   MCDP 6, 71; and ADRP 6-0, *Mission Command*, Change 2, March 2014, 2-15 to 2-17.

6   The AAR process is designated in SOP—*Reinforced Combat Brigade, Standard Operating Procedures*, Version 3.0.

7   *Commander 3rd Brigade Organisational* Intent, Headquarters, 3rd Brigade, Townsville, Australia, 2015.

8   Email response of Commander 3 Bde to Commander Combat Training Centre: *RFI to COMD CTC—Identify the Top Ten Training Challenges for Combat Team Commanders in 2015*, September 5, 15.

9   For example, there are significant differences in emphasis that are important and go to culture. A clear example is evident in a comparing USMC and US Army thinking. Marines emphasise control as feedback while the US Army acknowledges the feedback role of control but describes control as a "science" that must "regulate forces and direct the execution of operations to conform to their commander's intent." The net effect is that the army's is a more centralized vision of mission command.

10   *Commander 3rd Brigade Organisational Intent* 2015.

11   Marine Corps Doctrinal Publication 6, *Command and Control*, Washington, D.C.: Headquarters, United States Marine Corps, October 4, 1996, iii.

12   United States Marine Corps Doctrine, MCDP 6, *Command and Control*, 136.

13 *Defence Capability Plan 2012 {Public}*. The principal land C4I capabilities are characterised as Acquisition Category II, which describes projects that are major capital equipment acquisitions and are strategically significant to the ADF. They are characterized by significant project and schedule management complexities; high levels of technical, operating, and support difficulties; and complex commercial arrangements.

14 Eitan Shamir, *Transforming Command: The Pursuit of Mission Command in the US, British, and Israeli Armies*, Stanford University Press, Stanford, California, 2011, 22.

15 Carl von Clausewitz, *On War*, trans. and ed. Michael Howard and Peter Paret, Princeton, NJ: Princeton University Press, 1976, 119.

16 Australian Army Doctrine, Land Warfare Doctrine 1, *The Fundamentals of Land Power*, 2014 defines and details "army culture."

17 3rd Brigade Assessment Working Group and Decision Board 03/15, August 20, 2015, Sensitive: Personal.

18 3rd Brigade post-activity review (PAR) notes focused on this pragmatic positive characteristic for Operation Philippines Assist 2013-14 and Operation Pacific Assist 2015 in Vanuatu. Domestically, similar observations were made in response to Cyclone Ita and Marcia in north and central Queensland in 2014-15.

19 ANAO Audit Report No.1 2011–12, *The Australian Defence Force's Mechanisms for Learning from Operational Activities*, Canberra Jul7 2011, 19 para. 20

20 Australian Defence Force Doctrine, Executive Series, ADDP 00.1, *Command and Control*, 2-17 to 2-19.

21 Australian Army, Combat Training Centre Report, *The Status of Mission Command: Platoon to Formation Level 2005 to 2015*, correct as of September 29 2015. The CTC summary concludes there is "significant room for improvement across all mission command key elements from platoon to formation level." The CTC observes units across all formations of the army on a regular basis in near-mission conditions, warts and all. The CTC is a critical learning system and implement for corporate knowledge capture. Its unique organisational position and authority allows it to observe and record both strengths and weaknesses.

22 The "blue dots" refer to friendly force locations as appear on the Australian Army's Battle Management System (BMS) system screens. This capability to show such locations is roughly analogous to that in the US Army Blue Force Tracker system.

23 Australian Defence Force photo.

24 Norman Dixon, *The Psychology of Military Incompetence*, London: Pimlico, 1994, 176.

ACKNOWLEDGEMENTS

The authors thank the many—recognized and otherwise— granting interviews, submitting their thoughts, or otherwise supplementing our own thoughts regarding the application of mission command in the service of the Australian nation and those of partner nations worldwide. Special thanks to Roger Cirillo and Joe Craig at the Association of the United States Army for proposing the writing be undertaken and bringing the manuscript to its audiences. We also thank the following who variously provided supporting documents, read early drafts, or otherwise assisted us individually or collectively: Colonel (US Army, retired) Clint Ancker, Rebecca Constance, Dr. Howard Coombs, General Sir Peter Cosgrove, Dr. Rhys Crawley, Kay Dancey, Peter Edwards, Lieutenant General (US Army, retired) Donald Holder, Professor David Horner, Steve Keating, Dr. Craig Mantle, Professor Dan Marston, Lieutenant Colonel (Dutch Army, retired) Henk Oerlemans, Dr. Albert Palazzo, Ashley Rogge, Jennifer Sheehan, and Major General (Australian Army, retired) Mick Slater. Thanks as well to Ant Blumer, Doug Fraser, Mick Moon, Jim Simpson, and Dick Stanhope; who were key 1 RAR Group subunit commanders in 1993 Somalia, for their assistance in providing anecdotes and developing context for the Somalia chapter.

To the above we must add our spouses, partners, children, and friends who supported—or at least tolerated—the moments spent with heads cocked in concentration and fingers on keyboards in putting the following pages together.

AUTHOR BIOGRAPHIES

Dr. John Blaxland is Professor of International Security and Intelligence Studies at the Strategic and Defence Studies Centre at the Australian National University. Blaxland holds a Ph.D. in War Studies from the Royal Military College of Canada, an M.A. in History from the Australian National University, and a B.A. (Hons 1) from the University of New South Wales. A former Army Intelligence Corps officer, he is also a graduate of the Royal Thai Army Command & Staff College and the Royal Military College, Duntroon. In addition to a range of chapters and articles on intelligence, military history, and regional security issues, his publications include *A Geostrategic SWOT Analysis for Australia* (2019); *Tipping The Balance in Southeast Asia?* (2017); The Secret Cold War (2016); *East Timor Intervention* (2015); *The Protest Years* (2015); *The Australian Army From Whitlam to Howard* (2014); and *Strategic Cousins* (2006). He is also an editor for *In from the Cold: Reflections on Australia's Korean War* (2020) and *Niche Wars: Australia in Afghanistan and Iraq, 2001–2014* (forthcoming).

Lieutenant General John Caligari, AO, DSC, Australian Army (Retired) first deployed as a military observer with the United Nations Truce Supervision Organization in the Middle East before serving as a company commander and then operations officer of the 1st Battalion Group in Somalia as part of the US-led coalition Unified Task Force (UNITAF) in 1993. He commanded the 1st Battalion Amphibious Ready Element to the Solomon Islands for the evacuation of Australians during a period of civil unrest in 2000. Later that year, he commanded the 1st Battalion Group on operations with the United Nations Transitional Authority in East Timor. As Commander 3rd Brigade, he certified seven task groups ready for operations in Iraq and Afghanistan and then deployed as the Australian National Commander for Afghanistan in 2009. He is a graduate of the US Army Command and General Staff College and US Joint and Combined Land Component Commanders Course and has two master's degrees. He completed his military career as the Chief of Capability Development Group (Strategic J8) at Australian Defence Headquarters.

Dr. Peter J. Dean is chair of defence studies and director of the University of Western Australia's Defence and Security Program. Dean's major research areas include Australian strategic policy, the ANZUS alliance, and command, operations, and amphibious warfare. In 2014–15 he was the Department of Foreign Affairs and Trade Fulbright Fellow in US-Australian Alliance Studies at Georgetown University and in 2018, a Commonwealth Endeavour Research Fellow. An award-winning author, he is the primary author of nine books including *MacArthur's Coalition: US and Australian Military Operation in the Southwest Pacific 1942–1945* (2018) and (with Brendan Taylor and Stephan Frühling) *After American Primacy: Imagining the Future of Australia's Defence* (2019). He is the series editor of the Melbourne University Press Defence Studies Series, a former managing editor of the journal *Security Challenges*, and current board member of *Global War Studies* and *Australian Army Journal*.

Major General Chris Field serves in the Australian Army. He has commanded at each level from platoon, company, combat team, battalion, battle group, brigade, and joint task force, to include leading 36,000 people in the Australian Army's Forces Command. Combat deployments include East Timor, Iraq, and Afghanistan. He deployed twice on disaster recovery operations in Queensland, Australia, and on peacekeeping operations to the Middle East and Solomon Islands. He served as Vice Director of Operations, United States Central Command, as a Deputy Commanding General, 82nd Airborne Division, and as a planner with United States

Army Central. He is a graduate of the United States Army Land Component Commander Course, United States Marine Corps School of Advanced Warfighting, and United States Marine Corps Command and Staff College.

Dr. Meghan Fitzpatrick is a strategic analyst and an adjunct professor in War Studies at the Royal Military College of Canada. A graduate of King's College London, she is the author of numerous publications including *Invisible Scars: Mental Trauma and the Korean War* (University of British Columbia Press, 2017). Specializing in the history of operational stress injuries and military health, her work has appeared in distinguished journals such as Oxford University's *Social History of Medicine* and Taylor & Francis' *War & Society*.

Lieutenant General John J. Frewen, DSC, AM, Australian Army is a career infantry officer who specialized in rapid response forces. In 2003, as CO of the 2nd Battalion, Royal Australian Regiment (2 RAR), he led a multinational military intervention force supporting police to re-establish law and order in the Solomon Islands. This combined-joint task force comprised almost 1800 troops from five nations supporting a regional police effort. His other service includes deployments with the UN in Rwanda in 1994, NATO in Afghanistan in 2007, and command of all Australian forces across the Middle East in 2017. Lieutenant General Frewen's recent postings include that as Principal Deputy Director-General of the Australian Signals Directorate and Commander Defence Covid-19 Task Force.

Dr. Russell W. Glenn's US Army career included service in Korea, Germany, the United Kingdom, and a combat tour in Iraq with the 3rd Armored Division during Operations Desert Shield and Desert Storm. He served on the faculty of the Strategic and Defence Studies Centre at The Australian National University and with the United States Army Training and Doctrine Command after sixteen years in the think tank community. Dr. Glenn has degrees from the United States Military Academy, University of Southern California, Stanford University, US Army School of Advanced Military Studies, and University of Kansas. Past research includes studies on urban operations, counterinsurgency, military and police training, and intelligence operations. He is author of *Reading Athena's Dance Card: Men Against Fire in Vietnam*; *Rethinking Western Approaches to Counterinsurgency: Lessons from Post-Colonial Conflict*; and a forthcoming book addressing disasters in megacities.

Dr. Bob Hall graduated from the Royal Military College, Duntroon, in 1968 and served as an infantry platoon commander in the 8th Battalion the Royal Australian Regiment during its 1969–1970 tour in Vietnam. He is now a Visiting Fellow at the University of New South Wales Canberra. He is a military historian and currently leads the Military Operations Analysis Team in studies relating to Australia's involvement in post-1945 counterinsurgency operations. His publications include *Combat Battalion: The Eighth Battalion in Vietnam* and (with Andrew Ross and Amy Griffin) *The Search for Tactical Success in Vietnam: An Analysis of Australian Task Force Combat Operations*.

Brigadier Ian Langford, DSC and Bars, Australian Army is a career Special Forces infantry officer. He has commanded on operations at the platoon, company, task group, joint task force, and regimental levels. He has served on multiple occasions in East Timor, Solomon Islands, Bougainville, Afghanistan, the Southwest Pacific, Iraq, Israel, Syria, Lebanon, and on domestic counter terrorism duties. He is a graduate of the United States Marine Corps Command and Staff College and School of Advanced Warfghting.

Major General Roger Noble, DSC, AM, CSC, Australian Army is a 30-year soldier with a background in the Royal Australian

Armoured Corps. Through 1989 to 2004, Major General Noble served in a variety of regimental appointments in cavalry, APC, and tank units. He has an extensive staff background in capability development, concepts, and modernization. He completed his posting as Commander 3rd Brigade in Townsville in December 2015. Major General Noble has completed six operational tours of duty in Iraq, East Timor, and Afghanistan in a variety of command and staff appointments between 1992 and 2017. He was born in Cairns, Queensland, and is a keen surfer, ex-rugby player, and fisherman.

Dr. Peter Pedersen, AM was consultant historian for the Sir John Monash Centre and other Commonwealth commemorative projects on the Australian battlefields of the Western Front and for the ANZAC Museum in Beersheva, Israel, which commemorates the ANZAC campaign in Sinai and Palestine. A graduate of the Royal Military College, Duntroon, the Australian Command and Staff College, and the University of New South Wales, he commanded the 5th/7th Battalion, the Royal Australian Regiment, served as a political and strategic analyst at the Australian Office of National Assessments, and was Assistant Director at the Australian War Memorial. Dr. Pedersen's ten books include the acclaimed *Monash as Military Commander* and studies of the Gallipoli campaign and the battles of Fromelles, Villers-Bretonneux, and Hamel. He has presented many television and radio documentaries on Australia in the First World War, led numerous battlefield tours in Europe and Asia, including leading and organizing the first British tour to Dien Bien Phu, and appears frequently in the Australian media.

Major General Anthony Rawlins, DSC, AM, Australian Army has command experience as a troop leader, squadron commander, and commanding officer of the 2nd Cavalry Regiment. Recent staff appointments have included Colonel Plans at Headquarters 1st Division/Deployable Joint Force Headquarters, Director General

Military Strategic Commitments, and Deputy Chief of the Australian Army. He has served as a military observer with the United Nations Truce Supervision Organization in Israel and Lebanon (1999), commanding officer of Overwatch Battle Group West—Two in Iraq (2006–2007), and Chief Combined and Joint Operations (CJ3) at Headquarters International Security Assistance Force in Afghanistan (2014). His tertiary qualifications include bachelor's degrees in Arts and Law and master's degrees in Arts, Management, and Defence Studies.

Brigadier Chris R. Smith, DSC, CSC is the Australian Army Director General Land Operations (G3). He commanded 2 RAR Battle Group in Afghanistan in 2011, having in 2006 been operations officer for the same organization in Iraq. His operational experience also includes United Nations Assistance Mission for Rwanda 2 in 1995 where he served as a platoon commander and as an observer in the Middle East with the United Nations Truce Supervision Organization in 2002. Brigadier Smith is a graduate of the US Army's Command and General Staff College and School of Advanced Military Studies. Brigadier Smith holds a Bachelor of Arts, History from the University of New South Wales and a Master of Military Art and Science from the United States Army School of Advanced Military Studies.

www.ingramcontent.com/pod-product-compliance
Lightning Source LLC
Chambersburg PA
CBHW021547210326
41599CB00010B/341